The Digital (R)Evolution of Legal Discourse

Foundations in Language and Law

Edited by
Janet Giltrow
Dieter Stein

Volume 10

The Digital (R)Evolution of Legal Discourse

New Genres, Media, and Linguistic Practices

Edited by
Patrizia Anesa and Jan Engberg

DE GRUYTER
MOUTON

ISBN 978-3-11-221498-5
e-ISBN (PDF) 978-3-11-104878-9
e-ISBN (EPUB) 978-3-11-105036-2
ISSN 2627-3950

Library of Congress Control Number: 2023936705

Bibliographic information published by the Deutsche Nationalbibliothek
The Deutsche Nationalbibliothek lists this publication in the Deutsche Nationalbibliografie;
detailed bibliographic data are available on the internet at http://dnb.dnb.de.

Cover image: kokouu/E+/Getty Images
Typesetting: Integra Software Services Pvt. Ltd.
Printing and binding: CPI books GmbH, Leck

www.degruyter.com

Contents

Patrizia Anesa & Jan Engberg

Investigating legal discourse in the digital landscape

1 Exploring genres, media and discourse practices

Legal culture has increasingly felt the tension generated by digital transition, to the extent that scholars are urged to reflect upon how innovative technologies may harm the foundations of legal processes and practices, while, at the same time, offering new opportunities. Understanding how the digital revolution has affected language and discourse practices in the field of law is a demanding, though essential, task. Digitalization poses significant societal challenges in the application of the law, and this book aims to explore the complex nature of the techniques and discursive strategies which emerge in the relationship between the different stakeholders (including non-experts). By adopting a discourse analytical perspective, which combines both qualitative and quantitative approaches, the different contributions explore the hybridity of new legal genres, and communicative processes, by focusing on how digitalization can affect them.

This volume also provides a forum for discussion of the investigation of Legal Discourse from a digital humanities' perspective. Consequently, it attempts to create opportunities for integrating the work of linguists, computer scientists, and legal scholars, who focus on the analysis of the processes related to the digitalization of law, and its popularization in new genres and media, by adopting a profoundly interdisciplinary approach.

The volume is divided into three main parts, i.e.:
- Background and models in researching digital legal discourse
- Professional practice in the digital world
- Popularization and dissemination via digital tools

The first section is opened by Mary Lavissière's chapter, which focuses on the conceptualization of legal macrodivisions in legal documents. Using maritime contracts to illustrate these macrodivisions, there is a specific reference to how they have evolved in the digital era. The study takes, as a point of departure, the assumption that different types of users will be able to understand legal documents more easily if the documents' information structure is made more explicit. This is indeed a strong premise in a vast range of areas, including teaching legal language as a language for specific purposes (LSP), in the Plain Language movement, and in

https://doi.org/10.1515/9783111048789-001

legal design, as well as in researching legal language. Despite this claim, many of the basic macrodivisions in legal documents, such as titles, headings, and articles, have been largely undertheorized. In addition, other units, such as "moves" and "steps" are sometimes subject to unclear and contradictory definitions for legal documents. This chapter aims to offer a definition of these units in the context of the growing digitalization of legal texts and integrates them into a theory of legal genres. This contribution concludes with the relevant consideration that it takes an interdisciplinary approach to the definition and theorization of the macrodivisions of legal texts, including linguistic, natural language processing, LSP teaching/learning, and legal design perspectives.

The second chapter, "Processing of personal data in Court proceedings: a model for linguistic and legal studies", is by Laura Clemenzi, Francesca Fusco, Daniele Fusi, and Giulia Lombardi. It offers some key methodological considerations for the investigation of legal texts. In particular, it effectively tackles the issue of anonymization. Indeed, although Court proceedings, and in particular counsel documents, are a central resource for the investigation of legal discourse, the presence of personal data makes their collection difficult and their dissemination impossible. Consequently, they have not been systematically analyzed, and the problem of anonymizing texts has crucial implications for their use in research processes. The anonymization methods which are generally adopted in Italy and Europe are based on the obscuration of personal data, which generally takes place by erasing them with black strokes, or replacing them with asterisks, omissis, letters, or other graphic signs (Candrilli 2021). These practices, however, strongly affect the readability of the texts and often hamper the possibility of linguistic analysis. For example, erasing sensitive data such as anthroponyms, toponyms, or dates, makes it more difficult to identify and distinguish the different parties involved and to reconstruct the reported facts. For a comprehensive investigation (be it linguistically or legally focused), it is essential to access readable and complete texts. In this regard, the authors present an empirical project, involving jurists and linguists from four Italian universities (Genoa, Florence, Lecce, and Viterbo), which aims at creating a new resource for the effective writing of Court proceedings written by the defence, by building a synchronic corpus of about three million words and a searchable database (Gualdo and Clemenzi, 2021). This study is based on "pseudonymization", which ensures the full readability of the texts and the protection of the personal data contained therein. This process is defined as "the processing of personal data in such a manner that the personal data can no longer be attributed to a specific data subject without the use of additional information, provided that such additional information is kept separately and is subject to technical and organizational measures to ensure that the personal data are not attributed to an identified or identifiable natural person" (General Data Protection Regulation, Reg. EU 2016/679). The research team focused on the development of software which can automatically replace personal data with fictitious data of the same category extracted from predefined lists, after applying a light markup to the texts, which adds information, rather than loses it. Through this tool it is also possible to maintain constant and consistent replacements within the same text, or within several texts relating to the same judgment. The authors present a model of semi-automatic processing of personal data developed within the aforementioned project, showing how it can be useful for linguistic and legal studies on Court proceedings (or other legal texts), as well as for dissemination purposes.

The second part of the volume focuses more specifically on legal practice and its evolution in the digital era. Karin Luttermann and Alessandra Lombardi discuss how legal specialists narrate their stories on the web, and investigate the functions and the characteristics of professional biographies on Italian and German law firm websites. This chapter offers a comparative overview of the main pragmatic and textual features of legal professional bios from selected samples. It also provides a reflection on the interplay between global trends in online personal and professional branding, and culture-specific representations and attitudes towards the changing role of legal experts in the digital age. Finally, it discusses the relationship between the emergence of new business models in the international legal arena (e.g. 'boutique' law firms) and innovative forms of digital self-presentation which reflect changing social, cultural, and professional values across cultures, as well as including locally-specific features. In particular, this work focuses on a case study comparing self-presentation practices on boutique law firm websites in Italy and Germany, with a view to contributing to a comparison of epistemic cultures (i.e. the way specialist knowledge is constructed in a discipline on the basis of a common conceptual and value-oriented background; see Liebert 2016). In order to do so, the authors draw on the results of previous comparative studies on the pragmatic features of advertisements for legal jobs in different countries (see Luttermann/Engberg 2017 and Luttermann 2018). These show substantial homogeneity of communicative functions across cultures, with job adverts gradually evolving into instruments of employer branding (see Nielsen/Luttermann/Lévy-Tödter 2017), as well as culture-specific dimensions with regard to shared professional values and attitudes in different national contexts. The pragmatic analysis, centered on the sections of law firm websites where staff profiles are presented, investigates how professional image is shaped, communicated, and remediated in the examined legal cultures. The study assesses – from a cross-linguistic and cross-cultural perspective – the salience of professional biographies (as a new emerging digital subgenre of self-presentation vs. more traditional formats such as CVs), the presence of professional stereotyping (auto-, hetero- and meta-stereotypes disclosing specific cultural standards), and the impact of the digital medium (the website as a multimodal communication platform) on the personal marketing of legal specialists promoting their expertise and services on the web.

The reflection on the role of legal professionals in the digital era continues with Giuliana Diani's paper, "Stance features in legal blogging", which explores how law bloggers use stance resources to position themselves within the legal community. Drawing on Hyland's (2005a, 2005b) analysis of stance, a quantitative and qualitative analysis is presented based on a small corpus of blog posts written by law scholars commenting on legal cases relating to US and UK court decisions. The study combines corpus methodology with a discourse analytic approach and identifies stance features marking bloggers' presence in relation to their arguments and audience. The analysis shows that law bloggers prefer to downplay their commitment to the positions advanced in the posts and convey their authorial stance through boosters and attitude markers, rather than self-mention. This aspect appears in contrast to the distinctive feature generally associated with blogs, i.e. individualistic self-expression. These findings may contribute to a better understanding of the legal practices in web-mediated communication.

The theme of legal practice is also investigated in the following chapter, entitled "The Internet as a game changer in legal communication: Arbitration on the move". This is the result of a fruitful collaboration between Daniel Greineder and Dieter Stein, and their complementary backgrounds, respectively in law and linguistics, are put at the service of a profoundly multidisciplinary study. The chapter starts with the consideration that the advent of electronic communication and the Internet has given us a wealth of studies and volumes on medial-varietal linguistics. Main issues concern the properties – structural, communicational, pragmatic – of the new language medium, to what extent they are the same as, or different from, written and spoken media, and to what extent the new medium generates new genres (or to what extent genres and interactional modes persist unchanged or modified). The authors notice that what is discussed comparatively rarely are the effects on societal domains and their genres, such as the law, with their special constraints in terms of knowledge construction, management, transfer, and negotiation. This chapter takes up this strand of argumentation and considers the effect of the technical and pragmatic affordances of the Internet in a very specialized area, that of international arbitration. The first aspect considered is the legal status of arbitral awards, increasingly published online, which is resulting in a growing constraint on the arbitration process due to the sheer number of obtainable precedents. Another effect of the growth in the availability of published settlements is the rise of a doctrine of the "soft precedent" which, although not strictly legally binding, is clearly exerting an influence on decision-making, as well as having an effect on quantitative decision expectations thereby, effectively, changing the substance of the law. A further issue discussed is the physical condition of online video arbitration, the use of which was accelerated by the COVID-19 pandemic. Non-adapted paralinguistic behavior, with attendant uncertainties about what is communicated in a video setting, has led to an incipient trend of the taking of witness evidence by video link or a web-based platform, and these processes can have an impact on the way texts are produced and received.

The second section of the volume is concluded by Jekaterina Nikitina's chapter, "Interaction dynamics and remote participation in the International Criminal Court proceedings", which reflects on how digital practices, such as videoconferencing, can affect the conduction of proceedings. Indeed, even though remote hearings, and the respective digital effects on courtroom discourse, predate the pandemic by at least two decades, due to the COVID-19 situation they have become the 'new normal'. This has raised a number of ethically- and legally-challenging issues concerning the rights of defendants and witnesses to be heard in-person (International Commission of Jurists 2020: 6), and inviting research into the legilinguistic aspects of remote courtroom discourse. The hybrid jurisdiction of the International Criminal Court (ICC) relies on the adversarial model of proceedings, based on procedural fairness and the talk-in-interaction model (Licoppe 2021: 363). This study analyzes the trial stage of *The Prosecutor v. Alfred Yekatom and Patrice-Edouard Ngaïssona*, a case heard before the ICC where the defendants were accused of war crimes and crimes against humanity allegedly committed in the Central African Republic (CAR). The ICC has long used videoconferencing for some stages of proceedings, and the pandemic-related travel restrictions increased such use. In fact, in the Yekatom and Ngaïssona case, some of the participants in the hearing took part via video-link. This study explores the institutional interaction dynamics between in-person and remote participants in the trial, as well as the framing of remote participation. The study materials include trial transcripts in English and videos (floor / English booth). The main theoretical-methodological framework is that of Conversation Analysis (Sacks et al. 1974), which is supplemented by multimodal insights into legal interaction (Matoesian and Gilbert 2018; Licoppe 2021). The analysis draws on Goffman's interactionist notions (1981), the notion of participation framework (1974), along with deictic referential practices to address the issue of participation and agency, in a virtual court, of the defendants not able to be present in person (Licoppe 2021).

The final part of this volume tackles the complex issue of popularization and dissemination via digital tools, and is opened by Katia Peruzzo and Federica Scarpa with their chapter "TrIACLE: Making national immigration and asylum case law accessible to non-Italian web users". This study focuses on Italian case law developed through disputes concerning the rejection of applications for asylum or subsidiary protection by the Supreme Court of Cassation, i.e. the highest court of appeal in Italy. More specifically, it presents TrIACLE (Translated Immigration & Asylum Case Law in Europe), a translation project conducted at the University of Trieste whose underlying idea is that the monolingual tradition of Italian courts constitutes a limit to the dissemination of legal knowledge and the circulation of 'food for thought' beyond national boundaries. Therefore, TrIACLE's aim is to translate into English a selection of the most significant and recent judicial decisions of the Court of Cassation related to asylum and international protection, in order to make them available to a wider audience of non-Italian speakers. Although the standards applied in these decisions rely on EU and international law available in English, this type of translation is made significantly challenging by, on the one hand, the fact that the applicable national legislation in the cases under examination is in Italian and, on the other, by the peculiarities of Italian judicial drafting (Cortelazzo 1995; Ondelli 2014; Scarpa & Riley 1999). First, this contribution describes the interdisciplinary translation team in charge of the project, the methodological choices that it had to make, as well as the work processes which were followed in order to overcome the challenges posed by this type of inter-lingual translation. These issues are due not only to the culture-boundedness of the legal system underlying the source texts, but also to the often-convoluted nature of the language used by Italian judges. The second part of the chapter presents concrete examples to show the fundamental role of the preliminary intra-lingual translation performed by the team members in the pre-translation phase. Indeed, making national case law intelligible to an international, and not clearly defined, audience often requires various forms of simplification of the source texts, which are only possible through the close collaboration of field and language experts.

The topic of popularization is the object of the final chapter, by Farida Buniatova, which is based on a case study of the American legal TV drama series, "How to Get Away with Murder". The popularization of legal information can take place via different mass media, which are often available via digital platforms. TV series can represent an effective tool for the dissemination of information and can popularize knowledge pertaining to many important spheres of human life, including the legal one. From this perspective, not only do these series have an informative function, but they also shape people's perceptions about the law (Brown Graves 1999). This chapter is aimed at studying the image of a criminal defense attorney and university professor represented in the popular American TV series. The study provides a thorough analysis of the image of Professor Keating, the protagonist, who is both a prominent attorney dealing with cases of the utmost complexity, and a professor at the prestigious law school at a fictional Philadelphia university. The main issues investigated are whether the image depicted in the series represents a realistic picture of the life of law students in the US, and whether the image of Professor Keating corresponds to the real image of a university professor. This chapter reflects on the extent to which TV drama series can shape human expectations about truth and justice, legal procedure, and the image of lawyers. This leads to an important reflection on whether their content needs to be accurate and aimed at teaching law, or rather to only serve the purpose of entertaining viewers, suggesting a dream world of justice, fairness, and order.

2 Future challenges and potentialities

Digitalization clearly brings with it new ways of disseminating and consuming legal content. Such ways include a myriad of internet-based sources of legal information. This process clearly leads to the fast accessibility of information but, at the same time, it multiplies the risk of inaccurate or false information. The possibilities for controlling the adequacy of information, and preventing its uncontrolled proliferation, are rather limited. In this regard, considerable scholarly attention has been devoted to the need to examine context, either manually or automatically, although defining the criteria for accessibility is also particularly complex.

Digitalization generates representations of law through new and non-traditional approaches, which often imply an ample use of multimodal practices. This has crucial implications, especially in terms of the dissemination and popularization of legal content. Consequently, such content may appear more visually attractive and easier to access, and may contribute to enhancing legal consequences and knowledge, but it may also ultimately be subject to trivialization.

The introduction of digitalization in law processes is a fact which cannot be avoided. Clearly, it has enormous potential regarding the improvement of efficiency in law-making and law-enforcement activities, as well as in the realization of justice, the relationship between legal authorities and citizens, and scholarly research focusing on legal discourse. At the same time, the debate around this topic is intense as some point out that digitalization may bring with it a dehumanization of law and justice.

Clearly digital processes are not intended as a replacement for human activities, but rather as an inevitable condition which should be exploited fruitfully. The implementation of digital processes can be beneficially adopted in the drafting of texts, in their archiving, recording, certifying, etc. It can also favor the creation of platforms which can help citizens to access legal texts more quickly and easily; it can provide researchers with tools which can render their activity more efficient and accurate; and it can ultimately help improve the speed of the realization of justice. These objectives can potentially be reached without degenerating into uncontrolled practices or into the violation of ethical rules or, even worse, human rights. Of course, in order to prevent these risks, actions need to be taken, and the efficacy of digital tools and practices has to be monitored constantly, while the emic perspective of all the actors involved, be they experts or laypeople, are also considered.

The use of digital technologies in making and enforcing laws has often been greeted with skepticism, especially as regards the possibility of resorting to artificial intelligence for the automation of justice. Although the automatic assessment of legal action is no longer an impossible scenario, it may raise considerable concerns in terms of legality, and fairness, especially if used in isolation. Thus, the conduction of specific operations may be assisted by artificial intelligence tools, although these tools cannot be allowed to make procedural decisions, as the human factor is still the quintessence of law.

Of course, the COVID-19 pandemic accelerated the introduction of digital processes (e.g. face recognition systems, online hearings, etc.), but this has not led to an uncontrolled introduction of digital technologies in legal activity. The use of intelligent digital systems in this context needs to be further investigated in order to assess the real risks and the benefits of digitalization. Ultimately, it should be seen as the inevitable evolution of legal discourse, rather than as a threat to its purity.

References

Brown Graves, Sheryl. 1999. Television and prejudice reduction: When does television as a vicarious experience make a difference?, *Journal of Social Issues* 55 (4). 707–727.

Candrilli, Fernanda. 2021. Il progetto di archiviazione e anonimizzazione. In Riccardo Gualdo & Laura Clemenzi (eds.), *Atti Chiari. Chiarezza e concisione nella scrittura forense*, 19–29. Viterbo: Sette Città.

Cortelazzo, Michele. 1995. Lingua e diritto in Italia. Il punto di vista dei linguisti. In Leo Schena (ed.), *La lingua del diritto. Difficoltà traduttive. Applicazioni didattiche. Atti del primo convegno internazionale*, 5–6 ottobre 1995, 35–50. Milano: CISU.

Goffman, Erving. 1981. *Forms of Talk*. Oxford: Blackwell.

Goffman, Erving. 1974. *Frame analysis. An essay on the organization of experience*. New York: Harper & Row.

Gualdo, Riccardo & Laura Clemenzi (eds.). 2021. *Atti Chiari. Chiarezza e concisione nella scrittura forense*. Viterbo: Sette Città.

Hyland, Ken. 2005a. Stance and engagement: A model of interaction in academic discourse. *Discourse Studies* 7(2). 173–192. http://dx.doi.org/10.1177/1461445605050365.

Hyland, Ken. 2005b. *Metadiscourse: Exploring Interaction in Writing*. London: Continuum.

Licoppe, Christian. 2021. The politics of visuality and talk in French courtroom proceedings with video links and remote participants. *Journal of Pragmatics* 178, 363–377.

Liebert, Wolf-Andreas. 2016. Wissenskulturen. In Ludwig Jäger, Werner Holly, Peter Krapp & Samuel Weber (eds.), *Sprache – Kultur – Kommunikation / Language – Culture – Communication. Ein internationales Handbuch zu Linguistik als Kulturwissenschaft / An international handbook of linguistics as a cultural discipline*, 578–586. Berlin: De Gruyter Mouton.

Luttermann, Karin. 2018. Kommunikativ-funktionale Analyse von werbenden Gebrauchstexten in der Wirtschaft. In Kerstin Kazzazi, Karin Luttermann, Sabine Wahl & Thomas Fritz (eds.), *Worte über Wörter. FS Elke Ronneberger-Sibold*, 301–318. Tübingen: Stauffenburg Verlag.

Luttermann, Karin & Jan Engberg. 2017. Kulturkontrastive deutsch-dänische Textanalyse von sprachlichen Handlungen in juristischen Stellenanzeigen. In Martin Nielsen, Karin Luttermann & Magdalène Lévy-Tödter (eds.), *Stellenanzeigen als Instrument des Employer Branding in Europa. Interdisziplinäre und kontrastive Perspektiven*, 107–131. Wiesbaden: Springer VS.

Matoesian, Gregory & Kristin Gilbert. 2018. *Multimodal Conduct in the Law. Language, Gesture and Materiality in Legal Interaction*. Cambridge: Cambridge University Press.

Nielsen, Martin, Karin Luttermann & Magdalène Lévy-Tödter (eds.). 2018. *Stellenanzeigen als Instrument des Employer Branding in Europa. Interdisziplinäre und kontrastive Perspektiven*. Wiesbaden: Springer VS.

Ondelli, Stefano. 2014. Drafting court judgements in Italy: history, complexity and simplification. In Vijay. K. Bhatia, Giuliana Garzone, Rita Salvi, Girolamo Tessuto & Christopher Williams (eds.), *Language and Law in Professional Discourse: Issues and Perspectives*, 29–45. Cambridge: Cambridge Scholars Publishing.

Sacks, Harvey, Emanuel A. Schegloff & Gail Jefferson. 1974. A simplest systematics for the organization of turn-taking for conversation. *Language* 50 (4). 696–735. https://doi.org/10.2307/412243.

Scarpa, Federica & Alison Riley. 1999. *La traduzione della sentenza di common law in italiano*. Trieste: EUT Edizioni Università di Trieste.

Mary C. Lavissière
Contract macrodivisions and the digital age

1 Introduction

The digitalization of legal texts has allowed researchers to reconsider how they can identify, describe, and influence larger linguistic structures in these genres. In terms of identification and description, the massive increase in the number of texts in digital format has led to the development of natural language processing (NLP) tools, such as Blackstone (Hoadley 2019) and LexNLP (2018). These programs can automatically identify titles, headings, themes, and certain clauses of legal documents, making them more machine-readable. Regarding influence on discourse structure, researchers and professionals aiming for more accessible digital legal texts have taken a prescriptive attitude. For example, in the Plain Language movement (Marazatto Sparano 2020: 46; James & Moriarty 2020: 64), and in legal design (WorldCC, Passera & Haapio 2022), there is a strong premise that an explicit information structure helps users of legal language to more easily understand legal documents. Finally, a coherent relationship between titles/headings, and the content they summarize, is essential to avoid legal risks (Trosborg 1997: 64, 66). In sum, as legal documents become digitalized and accessible online, understanding the nature of the relationship between macrodivisions, such as headings and the text they introduce, and articles, becomes more crucial.

Despite the importance of these larger discourse units, few models have been developed to describe them. This is especially true for genres in the legal field. Furthermore, in the existing models, there is little agreement about the distinction between larger discourse units such as *macrostructures – superstructures* (Van Dijk 1980) and *moves – structural elements* (Biber, Connor & Upton 2007). Our study is a first step towards filling this gap with regards to contracts. We propose a new model of the relationship between heading[1] and article based on an empirical diachronic study of 61 maritime contracts and their 561 republications. We examine how the lexical content changes in relationship to the headings, and

---

1 We define headings as subtitles linked to sections of text, or other subtitles that appear after the main title of the document and are indicated by a visual distinction (different indent, spacing before/after, different formatting such as larger font, bold, italics, etc.). While main titles are also important, we do not research them in this study.

https://doi.org/10.1515/9783111048789-002

how our findings contribute to the literature on larger discourse structures in legal texts, and more specifically, in contracts.

This chapter is structured as follows: first, we review the literature on types of macrodivisions and the relationship between headings and articles in contracts. Second, we present a corpus of maritime agreements that we studied to better understand the link between headings and articles (Lavissière & Fedi 2022). We also expose the textometric methodology used to study the evolution of the articles' content over time. This corpus was chosen because few studies have exploited these documents (Lavissière & Fedi 2022) and all of the changes to the corpus, since the first filing of the contracts, have been made public by the United States government (Federal Maritime Commission 2022). Third, we present our analysis of the relationship between headings and articles. Fourth, we discuss the literature and propose a model for the relationship between these units. Finally, we conclude the study by discussing its limits and future perspectives.

2 Literature review

2.1 Diverging descriptions of macrodivisions

The literature regarding macrodivisions distinguishes two types of larger discourse structures that are relevant for the relationship between articles and headings. The first is centered on the notion of content; the second, on form. In the paragraphs that follow, we discuss the different terms related to this dichotomy. We then underline some of the incoherencies in the literature regarding these divisions, and highlight why it is necessary to clarify their status.

In the preface to his seminal work, Van Dijk (1980: V) makes a distinction between *macrostructures* and *superstructures*. He writes, "Macrostructures are higher-level semantic or conceptual structures that organize the 'local' microstructures of discourse, interaction, and their cognitive processing. They are distinguished from other global structures of a more schematic nature, which we call superstructures. These are, so to speak, the global 'form' of the macrostructural 'content.'" Biber, Connor & Upton (2007) make a comparable distinction between *moves* (Swales 1990), which they define as functional units of communication, and *structural elements*. For the latter, Biber, Connor & Upton (2007) adopt Crossley's (2007) distinction, in which structural elements are standardized elements that do not vary and whose meaning does not depend on the writer's intention. Biber, Connor & Upton (2007: 53) write,

> All of the letters in the direct mail corpus include text that strikes the reader as somehow 'different' than the text in the body of the letter. Things like the date, address information,

and even the signature and the signature footer have a very different function in the direct mail letter than the communicative functions served by the move types described above. Their functions, while important and in many respects required, are more structural in nature than communicative.

In a similar vein to Van Dijk (1980), Biber, Connor & Upton (2007) claim that structural elements are analogous to grammatical words: they are the 'mortar' that hold the bricks of a specific type of discourse together.

While the theoretical models described above seem to converge, empirical studies of specific texts make the distinction between semantic and conceptual macrostructures-moves and superstructures-structural elements less clear. For example, the results of Biber, Connor & Upton's (2007) move analysis of fundraising letters led the authors to reconsider the division between structural elements and moves. While certain structural elements are considered obligatory and static, others are optional and therefore have a questionable status. The authors (2007: 58) conclude,

> It could be argued that a more careful analysis of at least some of these [structural elements] may be warranted in future studies. Indeed, it could be argued that at least some of these elements should be viewed as moves in themselves, as they are functional units of text serving a specific purpose that adds to the persuasive nature of the letters. Textual choices within these structural elements, for example how to phrase the salutation, are actually quite significant and can be viewed as something beyond a standardized template.

In a different study focusing on a specific legal genre, Groom & Grieve (2019: 212) identify headings in patents as moves, and not as simple structural elements. In sum, various authors conclude that it is not always possible to make a distinction between form and content when studying headings and the text which they introduce.

2.2 Macrodivisions in contracts

The literature regarding the nature of headings in contracts is sparse (Lavissière & Fedi 2022) and, where it does exist, is inconclusive. It is mainly restricted to professionally-oriented manuals (Bugg 2016; Cartwright 2016; Chesler 2009), or stylistic studies (Richard 2021). In the two studies that do focus on the genre characteristics of contracts, Danet (1980) and Trosborg (1997), the status of headings is not fully clear. Danet (1980: 472) points to a dominance of structural elements in contractual language. She states, "the style is frozen, and matters are 'all form and no content' or nearly so" (Danet 1980: 472). This is in direct contradiction to Van Dijk (1980: 100), who claims that headings are macrostructures, and, as such, are linked to content and, not only to, form. Trosborg (1997: 58) also highlights the fixed nature of some contracts, pointing out that there are standard formats for them, and they "are often drawn up from existing material." However, she also claims that lawyers make certain choices about the structure and content of contracts (1997: 64), and that headings must be linked to the content they introduce in order to be legally enforceable (1997: 66). This description is more characteristic of macrostructures-moves. In essence, the current literature does not indicate whether headings in contracts are macrostructures-moves or are superstructures-structural elements.

2.3 Variation as a proxy for type of macrodivision

While the status of headings in contracts is unclear in the literature, a common theme that emerges is that variation serves as a proxy for determining the nature of macrodivisions in discourse. Macrostructures-moves are seen as variable; super-structures-structural elements are described as fixed. This representation, however, seems to omit the link between the form of contracts and the legal framework in which they exist. Trosborg (1997: 61) clearly identifies this link, claiming that "[. . .] legal discourse within the scope of language of the law can be described as for-mula-based communication in as regards form: the legal rules and their accumu-lated precedents function as a sort of matrix or mould, so that the form side of contracts is, in effect, determined normatively." From this perspective, if the legal framework is well defined and static, we would expect less variation in contractual language. This hypothesis would extend to headings in contracts. On the other hand, in domains of law where the framework is less clearly defined, or subject to more changes, we would expect more variation in contractual language.

However, for a study of the variation of macrostructures in contracts to be rigorous, it would be preferable to have a full set of a type of contract in syn-chrony and in diachrony. As we shall see in the next subsection, the trend to-wards digitalization and public access to legal documents makes this possible for the contracts that underpin the maritime shipping industry.

2.4 Digital library of maritime contracts – a resource for studying macrodivisions

While they are the subject of few linguistic (Lavissière & Fedi 2022) and legal studies (Corruble 2018), the maritime contracts which are filed with the Federal Maritime Commission (FMC) of the United States of America are made available to the public on the FMC's website (Federal Maritime Commission 2022). Federal law in the

United States makes filing all common ocean carrier contracts compulsory. Refiling is also obligatory following any changes. In this way, the digitalization process, and the push for public access to legal documents that has accompanied the Plain Language movement in many countries (Montolió Durán 2012), allows for a minute study of how macrodivisions in these contracts behave synchronically and diachronically and in relation to the content of the contracts. In addition, these contracts exist in a unique legal environment that is currently under scrutiny for its concentrated market structure, its effects on local ports, and inflation, among other issues (Lavissière & Fedi 2022; Fedi, Lavissière & Lavissière 2022).

Two interdisciplinary studies of maritime contracts that were published between 1973 and 2021 (Fedi, Lavissière & Lavissière 2022; Lavissière & Fedi 2022) have shown that more recent maritime contracts are associated with more technical terms, i.e., *slot, vessel, allocation, capacity*, whereas older contracts are linked to more business-oriented terms, e.g., *notice, member, effective, date* (Fedi, Lavissière & Lavissière 2022). The authors interpret this change as an effort by maritime shipping companies to adapt to increasing pressures from maritime authorities of different jurisdictions. The latter increasingly cite the negative effects of these contracts on both the market and other actors in the maritime sector (Merk, Kirsteinand & Salamitov 2018).

These studies, however, have not yet observed how the headings of maritime contracts behave in synchrony or diachrony. In addition, no study has identified how the headings behave in regard to any changes in lexical content, although Trosborg's (1997) claims about the necessary link between heading and article content, as well as Van Dijk's (1980) classification of headings as macrostructures which summarize content, would suggest headings vary with the lexical content of the articles.

2.5 Research questions

Given the gaps in the literature identified above, we formulated three research questions for this study:
1) Do headings in contracts remain "frozen" diachronically as Danet's (1980) description of contract language would predict?
2) If the lexical content of a type of contract is changing over time, as observed by Lavissière & Fedi (2022), do headings also vary as Trosborg (1997: 66) would predict?
3) In light of the answers to questions 1) and 2), what is the nature of macrodivisions in maritime contracts?

3 Corpus and methodology

In the following section, we present the corpus of maritime contracts and the methodology – textometry and a qualitative study – used to answer our research questions.

3.1 Corpus

The corpus used for this study included 61 original maritime contracts and all their republications for a total of 651 contracts. All the contracts were downloaded from the FMC's agreement library (Federal Maritime Commission 2022) between May 2021 and January 2022. The corpus therefore includes all the contracts dealing directly with the ocean transport of containers that were available during the corpus constitution period.[2] The contracts in the corpus were published between 1973 and 2022, a period of 48 years that includes the earliest contract available and the contracts published most recently.[3] Only the body of the original contracts was included. The table of contents, appendices, and signature pages were excluded. Table 1 (below) shows a summary of the different subtypes of agreements included in our corpus:

Table 1: Summary of different subtypes in the corpus.

Classification according to FMC	Number of agreements
Alliance	3
Conference	1
No specific secondary classification = none	23
Maritime Terminal	1
Joint Service	2
Rate Discussion Agreement (RDA)	9
Slot Chartering Agreement (SCA)	18
Vessel Sharing Agreement (VSA)	5
Total	**61**

---

2 We did not include contracts dealing mainly with maritime terminals, i.e., equipment discussion agreements, assessment agreements, terminal rate discussion agreements, marine terminal facilities agreements, marine terminal services agreements, marine terminal joint venture agreements, or MTO cooperative working agreements.

3 While space limitation does not permit us to include the table listing the metadata of all the contracts included in the corpus, it is available on request from the authors.

After conversion into .txt, the corpus was coded for effective date, type of contract according to the FMC's classification, contract number, and title. The articles were also coded for theme in the original publication, which allowed for the creation of subcorpora based on the theme of the article. To create the subcorpora, we used the software program Iramuteq (Ratinaud 2014). The subcorpora were created after the qualitative study on the changes in articles and headings.

3.2 Methodology

We used two methods to study the corpus and the subcorpora. First, we carried out a qualitative analysis of the changes in headings, namely, headings that appeared, changed, or disappeared. We made two observations of the full corpus: to begin, we studied each single contract and all its republications and used a spreadsheet to track whether the headings and articles had changed over time. Afterwards, we assigned a color code to each heading and, using a spreadsheet software program, created a representation of the pattern of articles a) according to time and b) according to the subtype of the contract as classified by the FMC.

Second, we analyzed the subcorpora with the textometric software Iramuteq (Ratinaud 2014). This software includes a descending hierarchical clustering (DHC) algorithm originally proposed by Reinert (1983; 1990) and adapted by Ratinaud & Marchand (2012) for larger textual corpora. Iramuteq (Ratinaud 2014) lemmatizes the corpus and divides it into lexical lemmata (*active forms*), grammatical elements (*supplementary forms*), and *hapax legomenon*. It then divides the corpus into text segments. For this study, we set the text segment length to 45 words, which corresponds to Hiltunen's (2001:56) estimation of the average length of sentences in legal texts. Using the chi-square metric (χ^2) and correspondence analysis (Le Roux & Rouanet 2005), the DHC calculates which lexical lemmata occur significantly in the same text segments. The algorithm forms classes and a dendrogram showing the order of class formation. The researcher can then interpret and name the classes. We carried out this analysis on all the frequently reoccurring contract articles. We also included a chi-square study of the relationship between the classes and time.

4 Results

Our study of the 590 maritime contract republications shows that only 179[4] republications include language changes. This amounts to approximately 30% of the changes. In the following section, we focus on the language changes concerning the macrodivisions in the contracts. First, we present the results of the qualitative study of the changes in headings over time, and according to the classification of the contracts by the FMC. Second, we present the results of the textometric study of the articles and their lexical behavior over time and according to subtype.

4.1 Changes in headings

4.1.1 Overview of headings in maritime contracts

The headings in the corpus as a whole can be divided into two profiles. One group appears in almost all the contracts, as seen in Table 2.[5]

Table 2: Most frequently appearing articles.

Theme	Number of agreements
Name	59
Purpose	61
Parties	60
Geographic scope	60
Overview of agreement authority	62
Officials of the agreement and delegation of authority	54
Membership	45
Voting	54
Duration and termination	60
Law and dispute resolution	50

A second group is more variable. While some articles appear frequently, in approximately a third of the contracts, others are much less prominent, for example *Filing agent*, a heading which only appears in two contracts. The full set of less frequently occurring headings is provided in Table A in the appendix.

---

4 The other changes are related to parties joining/leaving, or a change in name/headquarters' administrative details.
5 The colors in Table 2 will also be used in Figure 1 and 2.

4.1.2 Individual contracts over time

Our results show that within an individual contract, headings are remarkably stable over time. In our study of the 590 maritime contract republications, only seven headings were added, deleted, or otherwise changed.[6] Notably, three of the changes are found in the 2006 republication of the *Indamex APL Agreement*. Table 3 summarizes these changes.

Table 3: Summary of changes to headings and articles in the corpus.

Title	Original publication	Change	Theme of article	Effective date
Trans Pacific American Flag Berth Operators Agreement	1985	Heading & article 14 added (formerly subsection 5.4)	Service contracts	1988
Caribbean Shipowners Association	1986	Heading & article 8 added	Voting	1998
Indamex APL Agreement	2002	Heading & article 7 removed	Voting	2006
Indamex APL Agreement	2002	Heading & article 8 added	Compliance	2006
Indamex APL Agreement	2002	Heading & article 9 added	Separate identity	2006
Indamex APL Agreement	2002	Heading & article 10 added	Non assignment	2006
THE Alliance Agreement	date 2016	Heading & appendix B added	Contingency fund	2017

Excluding the few exceptions in Table 3, individual contracts in our corpus show overall stability in their headings over time. In general, it seems that the original organization of the themes in a contract, as represented by the headings and not necessarily by the lexical content of the articles, resists change, even in the face of new legislation.

4.1.3 Headings in the whole corpus over time

If we look at the distribution of the headings over time, we see that a core set of headings in Table 2[7] is generally stable over time. This is shown in Figure 1:

---

6 Other major macrostructure divisions included adding subsections to articles.
7 The color codes were given in Table 2.

Figure 1: Headings in maritime contracts organized diachronically.

The first six headings appear almost uniformly and in the same order: *name, purpose, parties, geographic scope, overview of agreement authority, officials of agreement and delegation of authority.* There are six exceptions: *United States Flag Far East Discussion Agreement* (1973), *Liberty Global Logistics SC Line Cooperative Working Agreement* (2013), *Hoegh Bahri General Cargo Middle East Space Charter Agreement* (2015), *THE Alliance Agreement* (2016), *OCEAN Alliance Agreement* (2016), *Puerto Nuevo Terminals LLC Cooperative Working Agreement* (2019), *Foundation Carrier Agreement* (2020), and *World Shipping Council Agreement* (2020). These exceptions, however, are spaced over time and do not show a diachronic trend. If we include the headings *membership, voting, duration* we can identify a core block of headings in maritime contracts: 33 contracts adhere to the pattern: *name, purpose, parties, geographic scope, overview of agreement authority, officials of agreement and delegation of authority, membership, voting, duration.* We henceforth abbreviate this block to NPPGOOMVD. There is not, however, an established diachronic trend to the articles that appear after the core NPPGOOMVD headings.

4.1.4 Headings in the whole corpus according to subtype

Given that the corpus includes different subtypes of contracts, ranging from contracts that frame the discussion of shipping rates to those that allow for companies to borrow space on vessels or to operate the vessels of other shipping lines, we might expect variation to be determined by the subtype of contract. However, apart for the core pattern NPPGOOMVD identified in Figure 1, no clear trend appears if the contracts are grouped according to their secondary classification by the FMC. This grouping is represented in Figure 2, again with the color coding used in Table 2 and according to the subtypes in Table 1.

While there is a similarity between two of the three contracts classified as alliances, no other trend emerges from the data. In sum, our study of the patterns of headings, both over time and according to the subtype of contract, reveals that there is a core set of headings, or themes, that appear uniformly across time and the subtype of maritime contracts. With the exception of the alliances subtype, there is no clear trend in the type of headings, or their organization, after the core set of headings that appear in the majority of the contracts. This result leads us to study the text of the article which the headings introduce, and to determine whether this text remains stable or not.

Figure 2: Headings in maritime contracts organized according to subtype.

4.2 Change in the article content over time

We restricted our study of the lexical content of the articles to two of the core headings: *purpose* and *overview of agreement authority*. The other articles from the core pattern were not included. This was partly due to space limitation, but also because some articles, such as *name* and *parties*, contain almost exactly the same text. For example, the article *name* usually only contains one sentence, and the article *parties* includes administrative information, such as the parties' name and postal addresses. These articles, as such, constitute non-changing text whereas the articles, *purpose* and *overview of agreement authority*, which are longer and vary, offer the possibility of studying statistical trends in their lexical items.

4.2.1 Purpose

The heading purpose and the text of the article are generally the second macrodivision of the maritime contracts. As seen in section 4.1, the heading *purpose* does not change according to the subtype of the contract. The text under the heading *purpose* is supposed to provide the reason for the formation of the contract. As such, we may expect some variation in the content of the article according to the subtype because the corpus contains different subtypes of maritime contracts. Given the fixedness of the heading, if we hypothesize that it is functioning as a macrostructure-move summarizing the text it introduces, we would not expect wide lexical variation.

In contrast, however, the DHC shows that there is a relatively wide range of topics dealt with in this article. The clustering algorithm results in five different classes (C): C1 *authorize agreement*; C2 *transport special cargo*; C3 *provide services*; C4 *engage in cooperation*; and C5 *exchange data*, as shown in Figure 3.

Figure 3 also shows that C1 and C3 represent the largest proportions of the corpus, 25.3% and 26.4% respectively. These two classes are also relatively common to the article if we look at the association of the classes with the different subtypes of contracts. Figure 4 shows this association as measured by the chi-square metric:

Crucially, some classes are very closely associated with one subtype of contract: *transport special cargo* (C2) with *rate discussion agreements* (RDA), and *exchange data* (C5) with contracts having no specific classification (None). This shows that the article associated with the heading *purpose* varies, yet none of this variation is reflected by the heading itself. In sum, the lexical items in this article vary, but the heading does not. In this way, the heading is acting more as a *superstructure-structural element* rather than a *macrostructure-move*.

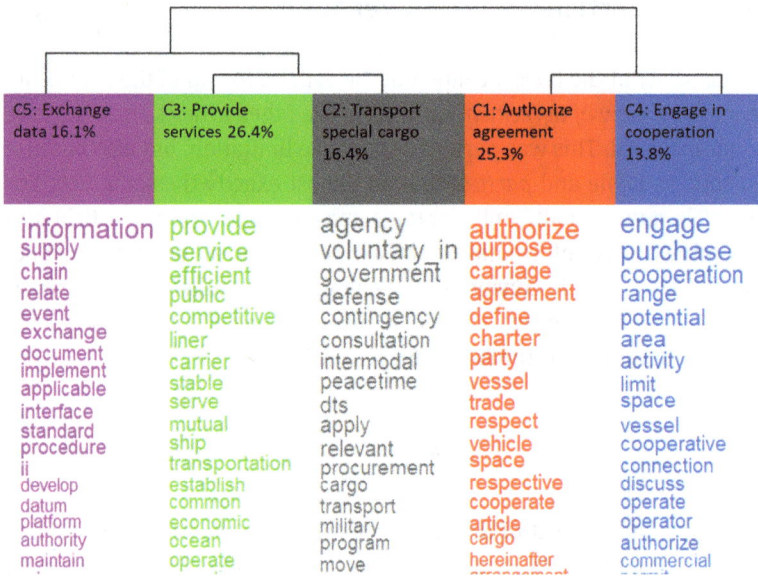

C5: Exchange data 16.1%	C3: Provide services 26.4%	C2: Transport special cargo 16.4%	C1: Authorize agreement 25.3%	C4: Engage in cooperation 13.8%
information supply chain relate event exchange document implement applicable interface standard procedure ii develop datum platform authority maintain	provide service efficient public competitive liner carrier stable serve mutual ship transportation establish common economic ocean operate	agency voluntary_in government defense contingency consultation intermodal peacetime dts apply relevant procurement cargo transport military program move	authorize purpose carriage agreement define charter party vessel trade respect vehicle space respective cooperate article cargo hereinafter	engage purchase cooperation range potential area activity limit space vessel cooperative connection discuss operate operator authorize commercial

Figure 3: DHC for the article *purpose.*

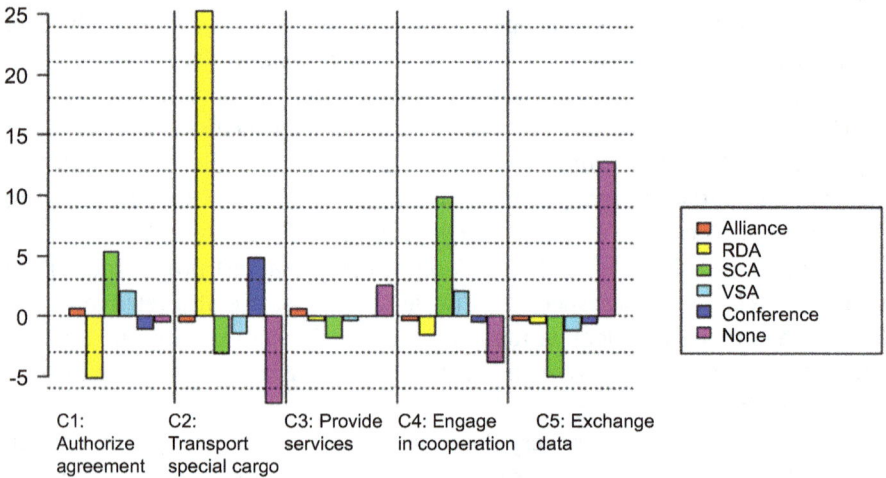

Figure 4: Association of the classes with the subtype of contract for the article *purpose.*

4.2.2 Overview of agreement authority

The article entitled *overview of agreement authority* is generally the fifth macrodivision in the maritime contracts. As with the other core macrodivisions, this heading does not vary over time or according to the subtype of contract. It is the longest of the core macrodivisions, with 50,242 tokens. This can be compared to the article entitled *purpose*, which has 3,750 tokens and the second longest article, *law and dispute resolution*, which has 13,882 tokens. *Overview of agreement authority* is arguably the most complex article as it is the article which authorizes the parties to collaborate in different ways. As with the article *purpose*, we expected some variation. However, if headings are macrostructures-moves, given this heading's fixedness, we would not expect wide variation in the lexical content of the text of the article itself.

Our results, however, show wide variation in the content of the article *overview of agreement authority*. The clustering algorithm represented in Figure 5 shows lexical worlds centered on: information exchange (C1); space (C2); shipping routes (C3); and relationships with third parties (C4).

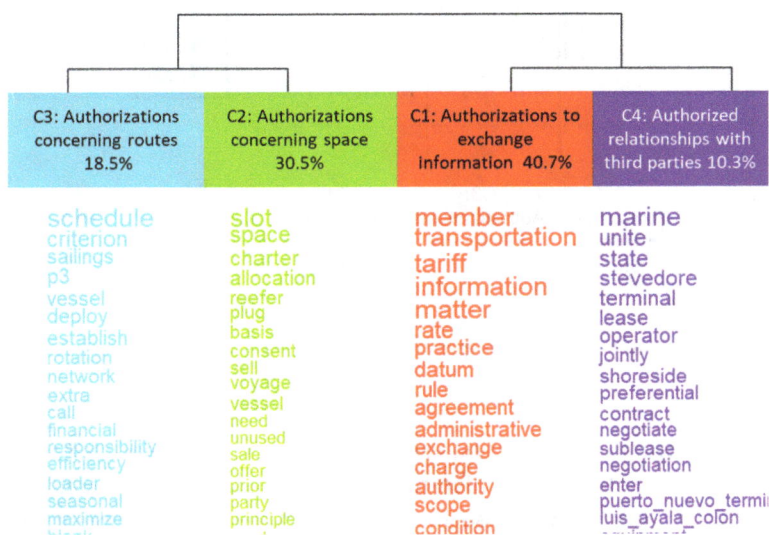

C3: Authorizations concerning routes 18.5%	C2: Authorizations concerning space 30.5%	C1: Authorizations to exchange information 40.7%	C4: Authorized relationships with third parties 10.3%
schedule	slot	member	marine
criterion	space	transportation	unite
sailings	charter	tariff	state
p3	allocation	information	stevedore
vessel	reefer	matter	terminal
deploy	plug	rate	lease
establish	basis	practice	operator
rotation	consent	datum	jointly
network	sell	rule	shoreside
extra	voyage	agreement	preferential
call	vessel	administrative	contract
financial	need	exchange	negotiate
responsibility	unused	charge	sublease
efficiency	sale	authority	negotiation
loader	offer	scope	enter
seasonal	prior	condition	puerto_nuevo_termi
maximize	party		luis_ayala_colon
	principle		

Figure 5: DHC for the article *overview of agreement authority*.

Authorizations regarding information exchange, whether they be about *tariffs, rates, datum, members,* or *practices,* are the most represented in the corpus, making up 40.7%. This is followed by authorizations concerning *space* on vessels (30.5%) and a set of lexical items concerning routes (18.5%). Authorizations for third party relationships only represent a small part of the lexical content of these contracts, presumably because different subtypes of contract deal with agreements between shipping lines and port terminals.

These different classes are not equally represented in the contracts. The classes are associated, first, with certain subtypes and, second, with certain time periods. We treat each of these associations separately. First, the differences are linked to specific subtypes of contracts, as seen in Figure 6.

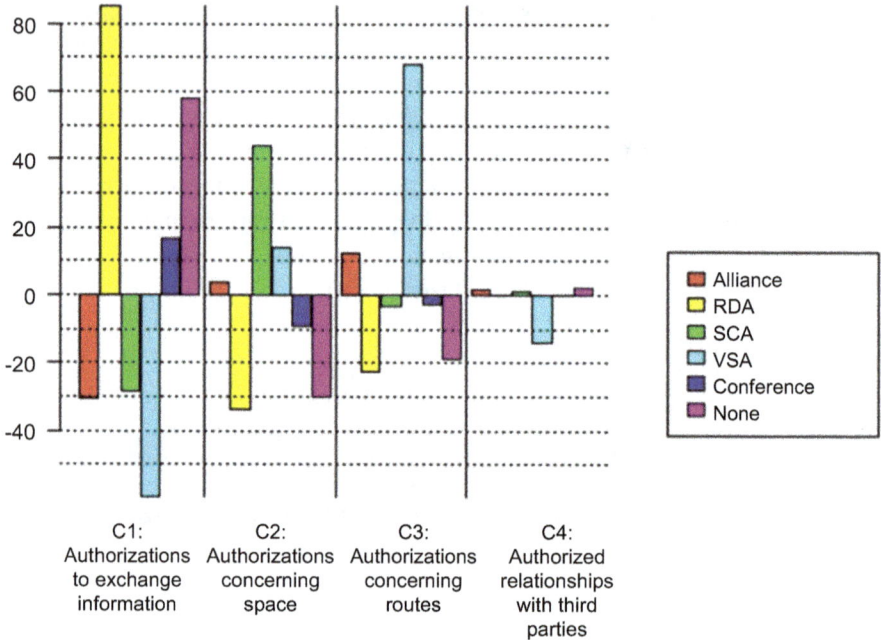

Figure 6: Association between classes and subtypes of contracts for the article *overview of agreement authority.*

As can be seen in Figure 6, information exchange (C1), is highly associated with rate-discussion agreements (RDA), conferences, and contracts with no secondary classification (None). Space (C2) is associated both with slot charter agreements (SCA) and vessel sharing agreements (VSA). Authorizations concerning routes are closely linked to vessel sharing agreements and alliances, which is to be expected as alliances are global vessel sharing contracts. Relationships with third parties are not strongly linked to these contracts, as they are dealt with in another type of contract.

Secondly, the classes are also associated differently over time. This relationship is represented in Figure 7:

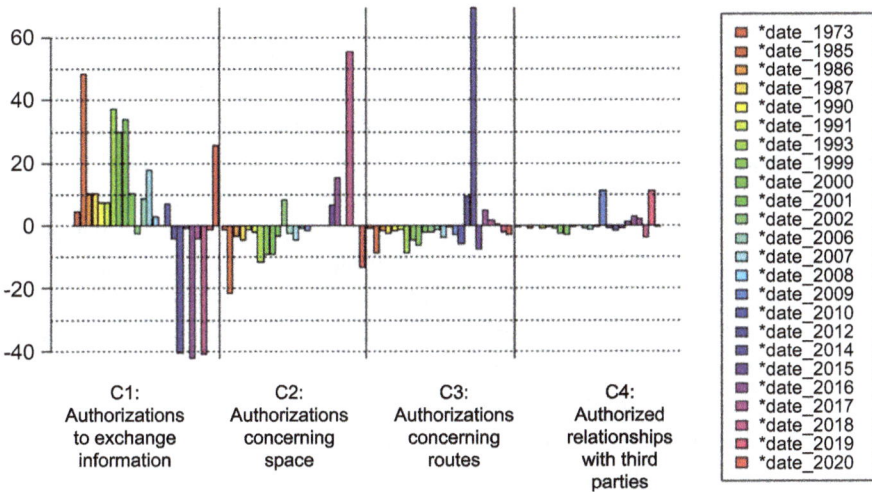

Figure 7: Association between classes and effective dates in the article *overview of agreement authority.*

Figure 7 shows that C1, authorizations to exchange different types of information, is more highly associated with older contracts than more recent ones. This trend is logical as the corpus includes rate discussion agreements and conferences which are no longer legal in many jurisdictions. In contrast, authorization concerning space sharing (C2) is more highly associated with recent contracts. The other two classes, which deal with route authorization (C3) and third-party relationships (C4) do not show a clear diachronic trend. This is presumably for two different reasons: first, the lexical words used to agree on route authorization seem to remain present in this article over time. Second, C4 does not make up a large part of the corpus because other subtypes deal with the business arrangements made with terminals.

5 Discussion

The gaps in the literature regarding headings and articles in contracts led us to formulate three research questions for this study. First, we asked if the headings of maritime contracts were indeed frozen diachronically, as Danet's (1980) description of contract language predicted. Our results showed that once a contract is established, the headings are very stable. Furthermore, from a corpus-wide perspective, we observed that a core group of six headings was very fixed and they did not vary according to their effective date or the contract subtype. Beyond the first six headings, a larger pattern (NPPGOOMVD) appeared in the majority of maritime contracts. In contrast, there was substantial variation in the headings that appeared after the core pattern. Other trends did not emerge, with the exception, perhaps, in the subtype *alliance*. As this subgroup only contains three contracts, it is difficult to make a strong conclusion about it.

Our second research question focused on the relationship between the heading and the text below it. Trosborg (1997) claims that the content of the text introduced by the heading must be semantically coherent with the heading itself. This was the case of some of the fixed headings in our corpus. For example, the headings *name* and *parties* did always reflect the semantics of their texts. Furthermore, the text of these articles did not change over time. On the other hand, more complex articles, such as *purpose* and *overview of agreement authority* varied either according to subtype and/or over time.

In sum, our results lead us to further question the soundness of the dichotomy macrostructure-move vs. superstructure-structural element. Our studies showed that there are at least three types of relationships between the headings and the text they introduce, as we see in Table 4:

Table 4: Typology of heading-article relationship.

Type	Heading	Text of article	Example	Nature of heading
1	Fixed	Changes in text are not reflected changes in heading	*Purpose, overview of agreement authority*	Superstructure – structural element
2	Fixed	Semantically coherent with heading	*Name, parties*	
3	Variable	Semantically coherent with heading	*Compliance, separate identity*	Macrostructure – moves

Type 1 corresponds to the description of a *superstructure-structural element* as it is stable across time and subtype. These headings do not closely summarize the text they introduce. It is possible that the headings appear solely to identify the document as conforming to the larger maritime contract genre rather than giving specific information about the text they introduce. While they fill a linguistic function at the text level, i.e. identifying a text as belonging to an established genre, they are problematic from a legal perspective. As Trosborg (1997) points out, headings must represent the text they introduce. Type 1 headings, however, are abstract enough to allow for substantial changes in the text they introduce without, themselves, changing. They are, in other words, relatively decoupled from their texts. The articles' texts, on the other hand, behave more as a *macrostructure-move*. They change their lexical content and, with it, their semantics, according to the needs of the parties and pressure from the legal "mold". While the latter is fragmented in regard to maritime contracts, it still imposes constraints on the parties, as seen in the 2014 decision by the Chinese authorities to reject a contract that would have constituted a major shipping line alliance (Fedi, Lavissière & Lavissière 2022; Lavissière & Fedi 2022).

Type 2 is problematic for the current dichotomy in the literature. Headings in this group are very stable and they appear in almost all the contracts. Their uniformity points to their being a necessary and stable part of the larger genre of maritime contracts. In this way, they are more like superstructures, according to Van Dijk (1980: 122):

> Superstructure categories appear to have a functional character: [. . .] they require specific information to be inserted in the category slots. [. . .] Heavier restrictions are required in institutionalized schemata: The psychological report, for instance, requires very specific information about subjects, experimental materials, design, and results of experiments. This is even more important in legal and institutional documents and forms, where the schematic categories are often questions that must be filled out with specific information (name, address, date of birth, profession, etc.).

In addition, unlike Type 1 headings, Type 2 are always strongly linked to the semantic content of their article. In this respect, they fit Van Dijk's (1980) definition of a macrostructure. In essence, they could fall into either category according to whether we rely on variation or semantic coherence as the criterion for characterization.

Type 3 headings vary over time but are closely linked to the text they introduce. For example, the heading *compliance* first appears in 2014, in contracts of the subtype *alliance*. Its appearance was clearly a rhetorical response to increasing pressure from competition law authorities who questioned whether these contracts allowed for hidden contracts, in violation of competition law (Lavissière & Fedi 2022; Fedi, Lavissière & Lavissière 2022). From this date, *compliance* appears in six other contracts as a heading and four as a subheading. Similarly, the article *separate identity*, which aims to confirm that the contracts are not hidden mergers, appears as an article in a 2014 contract of the subtype *space charter agreement*. It then appears in six other contracts as a heading and three as a subheading. In all these cases, the texts following the heading or subheading are semantically coherent with their headings. These headings and texts therefore function as macrostructures, giving a semantic theme to the reader of the text, or moves, communicating in various ways that the contracts conform to current competition law.

In short, as Table 4 shows, the nature of headings in maritime contracts does not fit into a single category if we adopt the criteria in the current literature. In other words, the criteria of a heading's fixedness, and its representativeness of the text that follows it, do not allow for absolute categorization. Fixedness is dependent on the definition of the scope of the corpus in terms of time period and number of subtypes included. It is also dependent on the level of institutionalization of the legal framework in which the genre functions, as Groom & Grieve's (2019) study of patents shows. This means that, given a small number of texts that belong to a very specific genre in a short period of time, it may be possible to distinguish which macrodivisions change and which are fixed. At the other extreme, given a very long period of time and many subtypes, it may also be possible to establish a very small group of stable macrodivisions. For example, a corpus of many different types of contracts in a common law legal system may show that the articles *name, parties, purpose* are what distinguish contracts from other legal genres. These cases show, however, that the nature of the macrodivisions would be determined mainly in function of the corpus and is not linked to an inherent difference in nature.

Rather than basing the categorization on actual variation alone, it may be interesting to categorize headings in terms of their potential for variation. The texts of some headings will have less potential for variation. This is the case of Type 2 headings such as *name*. As the potential for variation is low, the heading and its text function closely together. Type 3 has a high potential for variation as these are new headings and texts, which are created in response to an emerging need. This is the case of *compliance* and *separate identity*. In both Type 1 and Type 2, heading and text function as a highly coupled structure.

As opposed to the other types of headings, Type 3 headings show a different mode of functioning. The headings are fixed, but their text varies substantially. They are decoupled and should be analyzed as separate, but related, units. Type 3 headings have a superstructural role and fulfill the institutional community's expectations regarding the form of the contract. The text, however, varies significantly in terms of the parties' needs in the industry. Because the maritime industry is volatile (Caschili et al. 2014), and the legal framework is fragmented (Fedi, Lavissière & Lavissière 2022; Lavissière & Fedi 2022), the parties change the authorities granted through the contractual relationship, the governance structure, and other matters in response to evolving needs. In contrast, the headings remain stable as they have become part of the accepted institutionalized core pattern. This decoupling, however, also allows for what Bhatia (2008: 167) analyzed as language or genres for specific purposes that, "'bend' the norms and conventions." This bending is problematic from a legal point of view and should be analyzed in further studies.

6 Conclusion

In the present chapter, we have endeavored to show that the massive digitalization of, and public access to, legal genres allow for a new analysis of macrodivisions in legal texts. Because there is no agreement about the nature of headings in the literature, we aimed to confront the definitions with empirical data from a corpus of contracts. We used a publicly available corpus that documents all the changes in a specific type of contract, maritime agreements, to study the way that contractual macrodivisions behave over time. Our results showed that the current dichotomy of macrostructure-move or superstructure-structural element – which is based highly on the criteria of fixedness and relatedness between heading and article text – does not adequately categorize the way headings behave in maritime contracts.

We proposed a new typology of headings in maritime contracts, which includes three categories of headings. Rather than confirming a strict division between semantic or rhetorical types of divisions, such as macrostructures-moves, and formal superstructures-structural elements, our findings show that the status of macrodivisions is fluid. We also suggest that it is difficult to fully categorize these divisions because the categorization is highly corpus dependent. Finally, rather than basing the categorization solely on variation and the semantic relationship between the heading and its article, it seems important to include the criteria of the potential for variation between a heading and its article. It is also important to factor in the ability for professionals to decouple the text of the article from its heading to some extent.

Our findings and interpretations are, however, limited to the genre of maritime contracts. It will be important to extend this analysis to other types of contracts or legal documents to continue to understand the role played by macrodivisons in contracts and, more generally, other legal genres. Moreover, the link between microstructures, such as legal clauses or smaller linguistic units such as the sentence, should also be brought into the analysis of the way in which legal documents function in terms of macrodivisions. This project is underway, and we hope it will further clarify the functioning of legal macrodivisions.

Appendix

Table A: Less frequently occurring headings.

Theme	Number of agreements
Non assignment	18
Miscellaneous	16
Notices	13
Severability	11
Amendments	8
Separate identity	7
Compliance	7
Counterparts	6
Confidentiality	6
Language	5
Expenses dues	5
Independent action	4
Consultation and shippers' requests and complaints procedures	4
Agreement administration	3
Meetings and parliamentary procedures	3
Actions of members	3
Variation waiver	2
Filing agent	2
Relationships among non-shareholders	1
Trade association activities	1
Prohibited acts	1
Agreements with other carriers and persons	1
Performance of terminal and stevedoring activities	1
Financial	1

References

Bhatia, Vijay K. 2008. Genre analysis, ESP and professional practice. *English for Specific Purposes* 27(2). 161–174. https://doi.org/10.1016/j.esp.2007.07.005.

Biber, Douglas, Ulla Connor & Thomas Albin Upton. 2007. *Discourse on the move: using corpus analysis to describe discourse structure*. Amsterdam/Philadelphia: Benjamins.

Bugg, Stuart G. 2016. *Contracts in English: an introductory guide to understanding, using and developing "Anglo-American" style contracts*. München, Allemagne: Verlag C.H. Beck.

Cartwright, John. 2016. *Contract law: an introduction to the English law of contract for the civil lawyer*. Oxford/Portland: Hart Publishing.

Caschili, Simone, Francesca Medda, Francesco Parola & Claudio Ferrari. 2014. An Analysis of Shipping Agreements: The Cooperative Container Network. *Networks and Spatial Economics* 14(3). 357–377. https://doi.org/10.1007/s11067-014-9230-1.

Chesler, Susan M. 2009. Drafting Effective Contracts: How to Revise, Edit, and Use Form Agreements. *Business Law Today* 19. 35–37.

Corruble, Philippe. 2018. Les alliances mondiales entre le zist et le zest. *Le Droit Maritime Français* 807 (November-December). 867–878.

Crossley, Scott. 2007. A chronotopic approach to genre analysis: An exploratory study. *English for Specific Purposes* 26(1). 4–24. https://doi.org/10.1016/j.esp.2005.10.004.

Danet, Brenda. 1980. Language in the Legal Process. *Law & Society Review* 14(3). 445–564.

Federal Maritime Commission. 2022. Federal Maritime Commission Agreement Library. https://www2.fmc.gov/FMC.Agreements.Web/Public. (30 August, 2021).

Fedi, Laurent, Mary C. Lavissière & Alexandre Lavissière. 2022. Taking maritime shipping agreements out of the box. *Maritime Policy & Management*. 1–18. https://doi.org/10.1080/03088839.2022. 2160498.

Groom, Nicholas &Jack Grieve. 2019. The evolution of a legal genre. In Teresa Fanego & Paula Rodriguez-Puente (eds.), *Corpus-based Research on Variation in English Legal Discourse*. Amsterdam/Philadelphia: Benjamins.

Hiltunen, Risto. 2001. "Some Syntactic Properties of English Law Language": Twenty-five years after Gustafsson. In Keith Battarbee, Matti Peikola, Sanna-Kaisa Tanskanen & Risto Hiltunen (eds.), *English in Zigs and Zags: A Festschrift for Marita Gustafsson*, 56–63. Turku: University of Turku.

Hoadley, Daniel. 2019. Blackstone. Python. ICLR&D. https://github.com/ICLRandD/Blackstone. (17 October, 2021).

James, Neil & Greg Moriarty. 2020. Access starts with the precedent: Evaluating the language of leases. *Clarity* 81. 34–38.

Lavissière, Mary C. & Laurent Fedi. 2022. Maritime Cooperative Working Agreements: variability as a proxy for legal atomization. *Fachsprache* 44 (3–4): 130–47. https://doi.org/10.24989/fs.v44i3-4. 2029.

Le Roux, Brigitte & Henry Rouanet. 2005. *Geometric Data Analysis: From Correspondence Analysis to Structured Data Analysis*. Dordrecht: Springer Netherlands. https://doi.org/10.1007/1-4020-2236-0.

Marazatto Sparano, Romina. 2020. Plain language principles. *The Clarity Journal* 82. 43–47.

Merk, Olaf, Lucie Kirsteinand & Filip Salamitov. 2018. *The Impact of Alliances in Container Shipping*. International Transport Forum. https://www.itf-oecd.org/sites/default/files/docs/impact-alliances-container-shipping.pdf. (9 December, 2020).

Montolió Durán, Estrella (ed.). 2012. *Hacia la modernización del discurso jurídico: contribuciones a la I Jornada sobre la modernización del discurso jurídico español*. Barcelona, España: Universitat de Barcelona, Publicacions i Edicions.

Ratinaud, Pierre. 2014. IRaMuTeQ : Interface de R pour les Analyses Multidimensionnelles de Textes et de Questionnaires. Windows, GNU/Linux, Mac OS X. http://www.iramuteq.org.

Ratinaud, Pierre & Pascal Marchand. 2012. Application de la méthode ALCESTE à de "gros" corpus et stabilité des "mondes lexicaux": analyse du "CableGate" avec IRaMuTeQ. *Actes des 11eme Journées internationales d'Analyse statistique des Données Textuelles* 835–844.

Reinert, Max. 1983. Une méthode de classification descendante hiérarchique : application à l'analyse lexicale par contexte. *Les cahiers de l'analyse des données* 8(2). 187–198.

Reinert, Max. 1990. Alceste une méthodologie d'analyse des données textuelles et une application: Aurelia De Gerard De Nerval. *Bulletin of Sociological Methodology/Bulletin de Méthodologie Sociologique*. SAGE Publications Ltd 26(1). 24–54. https://doi.org/10.1177/075910639002600103.

Richard, Isabelle. 2021. La 'dé-spécialisation'de l'anglais juridique: exemple de l'évolution de la stylistique contractuelle en Common Law. In *Langues et langages juridiques* (Colloques & Essais),

143–159. Institut Francophone pour la Justice et la Démocratie. https://www.lgdj.fr/langues-et-langages-juridiques-9782370323064.html. (3 June, 2022).

Swales, John M. 1990. *Genre analysis: English in academic and research settings*. Cambridge: Cambridge University Press.

Trosborg, Anna. 1997. *Rhetorical strategies in legal language: discourse analysis of statutes and contracts* (Tübinger Beiträge Zur Linguistik 424). Tübingen: Gunter Narr Verlag.

Van Dijk, Teun Adrianus. 1980. *Macrostructures: an interdisciplinary study of global structures in discourse, interaction, and cognition*. Hillsdale: L. Erlbaum.

WorldCC, Stefania Passera & Helena Haapio. 2022. Skimmable headings. *WorldCC Contract Design Pattern Library*. http://contract-design.worldcc.com/skimmable-headings. (12 September, 2022).

2018. LexNLP. HTML. LexPredict. https://github.com/LexPredict/lexpredict-lexnlp. (17 October, 2021).

Laura Clemenzi, Francesca Fusco, Daniele Fusi & Giulia Lombardi

Processing of personal data in Court Proceedings: A model for linguistic and legal studies

1 A project for clarity in Court proceedings

Although Court proceedings are an important resource for linguistic and legal studies,[1] they have not been systematically analysed so far, as the presence of personal data, in particular in documents written by defence counsels, makes their collection difficult and their dissemination impossible.[2]

The issue has been addressed within the Research Project of National Relevance (PRIN) 2017 "La chiarezza degli atti del processo (AttiChiari): una base di dati inedita per lo studioso e il cittadino / Clarity in Court Proceedings (ClarAct): a new database for scholars and citizens", a project that involves both linguists and jurists from the Italian universities of Genoa, Florence, Lecce and Viterbo.[3] The project aims at creating a new resource for effective writing of Court proceedings drafted by the defence, by building a synchronic corpus of approximately three million words and a searchable database.

The specific objectives of the project are: firstly, the collection, for the first time in Italy, of a database of documents of the parties in the case, both civil and criminal, relating to proceedings pertaining both to the Supreme Court and to a set of geographically diversified Courts and Courts of Appeal; secondly, the qualitative study of both the textual and stylistic features and the rhetorical and argumentative structure of these texts, hitherto inaccessible to the research community;[4] and lastly, the

---

1 The text has been agreed upon and revised by all authors; however, for the sake of authorship, paragraphs 3 and 8 are attributed to Laura Clemenzi, paragraphs 1 and 4 to Francesca Fusco, paragraphs 2 and 5 to Giulia Lombardi, and paragraphs 6 and 7 to Daniele Fusi. We thank Jacqueline Visconti for her careful reading and valuable advice.
2 Early studies on the Italian language of the trial include Mortara Garavelli 2003a and 2003b, Sabatini 2003, Cavallone 2012, Caponi 2016, Dell'Anna 2016, Gualdo and Dell'Anna 2016.
3 The project was approved and funded by the Italian Ministry of Education, Universities and Research (MIUR) with the protocol number 2017BSECYX. On the relevance of clarity in Court proceedings, cf. Gualdo 2021 and Visconti 2018, 2021.
4 The collected documents are characterised by being produced in the interest of the parties and by being provided to researchers by the parties' representatives on an exclusively voluntary basis, in contrast to the collections of Court decisions, which are to a large extent public and available in archives and publications.

https://doi.org/10.1515/9783111048789-003

identification of a selection of particularly well written and well-constructed texts, collected in a section of the database, available to both private lawyers and judges as examples of good practice in clear and effective writing, and to citizens as an aid to the understanding of texts intended for them, but which are too often indecipherable, even for educated readers.[5]

2 Italian Court proceedings

The Court proceedings collected within the corpus AttiChiari are mainly defensive acts, i.e. texts drafted by lawyers in order to persuade the judge, and to argue, prove and plead the defendant's case (Alpa 2003; Mariani Marini 2003); Mortara Garavelli (2001: 19–34), and includes them within the category of judicial applicative texts. The interpretation of these texts tends to be binding (Sabatini 2003) and their textual typology is generally composite: there are narrative and expository texts traits in the exordium, argumentative texts traits in the motivation, and finally prescriptive texts traits in the preliminary enquiries (Visconti 2018, 2022). Moreover, compared to other legal texts, Court proceedings, especially those drafted by defence counsels, are characterised by a prevalence of connotation vs denotation, and by a greater personalisation (Gualdo and Dell'Anna 2016).

More in detail, the discourse organisation is functional to the lawyer's argumentative techniques; for this reason, the complexity of the reasoning is mostly manifested in chains of subordinate propositions and protasis often placed before the main proposition. Finally, there are many exclamatory and interrogative propositions, as an expression of that "dialogic mode" (Gualdo and Dell'Anna 2016: 630) that reproduces in writing the typical dynamics of the hearing with the judge.

The lexicon is heterogeneous and represents the wide variety of topics covered in the proceedings: we find technicalities, statutory lexicon, common words, terms derived from other branches, Latinisms and foreign words.

---

5 For more details on the objectives of the project and the procedures adopted, cf. the papers collected in Gualdo and Clemenzi 2021. For some examples of linguistic phenomena searchable in the AttiChiari database, see in particular Clemenzi 2021; other early studies on the corpus are in Dell'Anna (in press). Further information also regarding the composition and activities of the research group is available on the official project website, https://attichiari.unige.it.

3 Processing texts and personal data

The texts collected are also characterised by containing personal data, the disclosure of which would violate the right to confidentiality of the parties and the attorneys involved in the proceedings. Since the documents are provided by the lawyers on a voluntary basis, in order to obtain and analyse them, it has been necessary to devise a tool that would erase anthroponyms, toponyms, dates and any other personal data that would allow the recognition of the parties involved in the proceedings and the reconstruction of the reported facts.

In Italy, anonymisation practices are often used to erase personal data from the texts of judgments or other texts containing personal information, before they are made public. When dealing with very sensitive subjects (as the ones that involve minors) or when so-called "sensitive data" occur in the text, the attention to privacy has to be even greater.[6]

Traditional anonymisation practices consist of simply removing data by erasing them with black traits, by replacing them with asterisks, *omissis*, letters, or other graphic signs, or by leaving blank spaces, as in the following examples (Figure 1).[7]

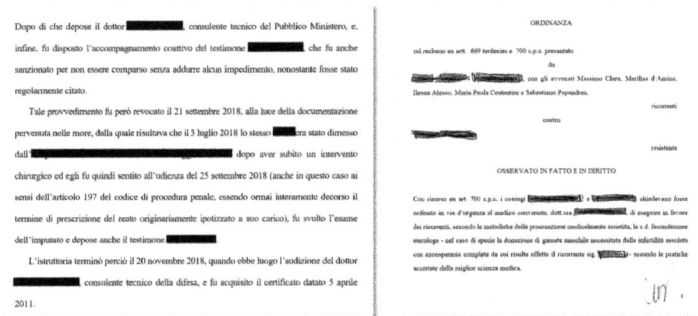

Figure 1: Examples of data anonymisation with black or pen traits in Italian documents (Candrilli 2021: 21).

---

6 The terms "personal data" and "sensitive data" are defined in the Reg. EU 2016/679, arts. 4 and 9.
7 In the following examples from Candrilli 2021, reference will be made to the publication of judgments. This will allow us to emphasise how the approach chosen in the AttiChiari project for the anonymisation of party procedural documents is characterised by an even greater exclusion of personal data from the texts than is usually the case. On Italian judgments, cf. Gualdo 2011: 453–462 and Santulli 2008.

Frequently, the visibility of elements such as the birth years of the parties and other chronological references (Figure 2) is preserved, also outside the Italian context (Figure 3).[8]

> Nel caso in esame i soggetti ricorrenti soddisfano i requisiti richiesti. La ricorrente, infatti, risulta, come da documentazione medica in atti, affetta da accertata sterilità secondaria da menopausa precoce. Essi, inoltre, sono maggiorenni, di sesso diverso, coniugati nel 2005 (come da certificato in atti) e in età potenzialmente fertile ▮▮▮▮▮ è del '74, mentre ▮▮▮▮▮ i è del '71).
>
> Le problematiche mediche, presenti nel caso in oggetto (infertilità risolvibile solo attraverso il ricorso alla fecondazione di tipo eterologo per come pacificamente emerge dalla documentazione in atti), non possono, però, essere risolte alla luce della disciplina dettata dalla l. 40 del 2004 atteso il divieto posto dall'art. 4, comma 3 della legge in esame.

Figure 2: Example of anonymisation not extended to chronological data in Italian documents (Candrilli 2021: 22).

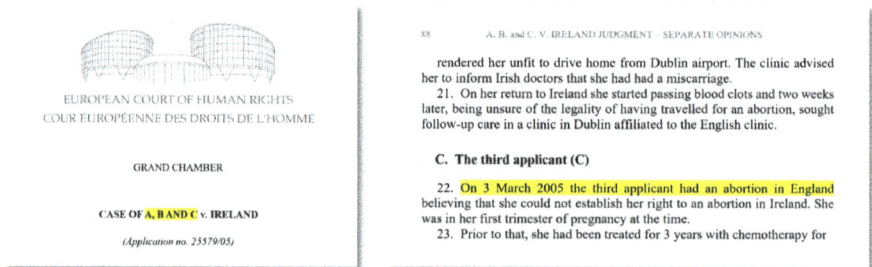

Figure 3: Example of anonymisation in the text of a judgment of the European Court of Human Rights (Candrilli 2021: 22–23).

These practices, however, affect the readability of the documents and hinder the possibility of linguistic analysis: erasing anthroponyms, toponyms, dates and any other personal data makes it impossible to identify and distinguish the different parties involved in the proceedings and to reconstruct the reported facts.

Readable and complete texts are instead crucial for both linguistic and legal studies, which aim at fully analysing the strategies used by lawyers in referring

---

8 As Candrilli 2021: 23 notes, in Figure 3 the names of the persons are replaced by letters (A, B, C, highlighting introduced by Candrilli), but the rest of the spatio-temporal references are left intact (this is a choice of method often adopted also in Italian judicial measures). You can read, e.g., «On 3 March 2005 the third applicant had an abortion in England».

to the assisted party, to the counterpart and to third parties involved in the trial; these questions become even more relevant when analysing, from an intertextual perspective, several documents related to the same judgement (Fusco 2021).

Therefore, in order to ensure, at the same time, the full readability of the texts and the protection of the personal data contained therein, the AttiChiari project has chosen to process the collected acts with a pseudonymization technique.

Pseudonymization is "the step where a pseudonym or code is added to [. . .] de-identified data" (Elger et al. 2010: 233). It is defined by the General Data Protection Regulation (Reg. EU 2016/679, art. 4 n. 5) as "the processing of personal data in such a manner that the personal data can no longer be attributed to a specific data subject without the use of additional information, provided that such additional information is kept separately and is subject to technical and organisational measures to ensure that the personal data are not attributed to an identified or identifiable natural person".[9] In particular, the research team developed a software which automatically replaces personal data with fictional data of the same category extracted from predefined lists, after applying a light markup to the texts.[10] This software also makes it possible to keep replacements constant and consistent within the same text, or within several texts relating to the same judgement.

For the purposes of linguistic analysis, it is fundamental to maintain the conceptual-semantic coherence between the original data and the fictional data, and the morphosyntactic coherence of such data with the context; it is important, therefore, that the new data univocally corresponds to the original one in all its occurrences in the text and that it preserves its gender, so as not to alter the morphosyntax of the sentence in which it is inserted. To this end, an automatic substitution of personal data, by means of a program that draws on predefined lists for names and modifies alphanumeric numerical sequences (such as dates, number plates, faxes, telephone numbers, etc.) proves to be the most effective solution, capable of reducing the risk of error and guaranteeing a uniform result.

Furthermore, since the study to be conducted on the texts is not only linguistic, but also legal, different sets of metadata should be envisaged according to the purposes: if, for example, linguistic analysis requires the insertion of metadata

---

9 The technique of "pseudonymization" is referred to in several parts of Reg. EU 2016/679 precisely as a measure that offers "appropriate safeguards" (art. 6(4e); see also arts. 25(1), 32(1), 40(2), 89(1) and whereas 26, 28, 29, 75, 78, 85, 156). Cf. Fusco 2021: 30–31.

10 In this paper the terms *mark(up)* and *annotation* are used as synonyms, even if in corpus linguistics *markup* indicates the encoding of contextual metadata relating to texts and of information relating to the structure and formatting of texts, while *annotation* indicates the practice of adding linguistic information. Cf. Freddi 2019: 19, Rodríguez-Puente, Blanco-García and Tamaredo 2019: 65–68.

relating to the para-text, legal analysis requires that the substitution of dates does not affect the chronological reconstruction of facts.

This paper aims at presenting the model of semi-automatic processing of personal data developed by the AttiChiari project, showing how it can be useful for linguistic and legal studies on Court proceedings (or other legal texts), also for dissemination purposes.

4 Annotation and pseudonymization

In order to fulfil the needs of the project, a new method for anonymising personal data has been devised, namely a semi-automatic annotation method inspired by the models of Douglass et al. 2004, Noumeir, Lemay and Lina 2007, Elger et al. 2010 and Dalianis 2019 for the pseudonymization of medical records, and Oksanen et al. 2019 for the pseudonymization of Finnish Court documents.

The annotation process involves a manual annotation followed by an automatic annotation (Fusi 2021). First of all, the operator manually annotates the source text directly in a word processing application by applying an agreed syntax, which indicates both the category of the personal data and the genre: the portion of text to be replaced is inserted in curly brackets, preceded by a tag indicating its category (anthroponym, toponym, date, etc.), and followed by a colon, on a {tag: text} pattern, as in the example shown in Figure 4.[11]

In particular, as shown in Figure 4, the tags used in the AttiChiari project to identify and replace personal data in the collected proceedings are:

a-f-f (*anthroponym, female, first*) for female anthroponyms;
a-m-f (*anthroponym, male, first*) for male anthroponyms;
a-l (*anthroponym, last*) for surnames;
j-f (*juridical person, female*) for legal entities of feminine grammatical gender;
j-m (*juridical person, male*) for legal entities of masculine grammatical gender;
t (*toponym*) for toponyms;
ad (*address*) for addresses;
m (*e-mail*) for e-mail addresses;
d (*date*) for dates;
n (*number*) for digits (e.g. telephone numbers, amount of money, land parcels, etc.);

---

11 The text reproduced in Figure 4 is extracted from a Court proceeding in which the original personal data have been replaced manually with fictional data for privacy reasons.

TRIBUNALE DI {t:ROVIGO}

R.G. N. {n:1234}/{d:2015}

PER

La **{j-f:Prima} Spa, in persona del legale rappresentante {f-lat:pro tempore}, Sig. {a-m-f:Mario} {a-l:Rossi}**, con sede in {t:Rovigo}, Via {ad:Giuseppe Garibaldi n. 23}, P.IVA {n:01234567890}, rappresentata e difesa dall'Avv. {a-f-f:Ada} {a-l:Verdi} (C.F. {u:VRDDAA67A41H620P}, fax {n:0425/123456}, pec {m:adaverdi@pec.it}) presso il cui Studio sito in {t:Rovigo}, Via {ad:Guglielmo Marconi n. 55}, ha eletto domicilio giusta delega posta in calce al presente atto

CONTRO

Il Sig. **{a-l:Grossi} {a-m-f:Zeno}** con l'Avv. {a-f-f:Alma} {a-l:Piccoli}

COMPARSA DI COSTITUZIONE E RISPOSTA

Con atto di citazione notificato in data {d:10/10/2014}, il Sig. {a-l:Grossi} {a-m-f:Zeno} conveniva in giudizio dinanzi l'intestato Tribunale la {j-f:Prima} Spa per sentire accertare e dichiarare la responsabilità della società convenuta per i vizi di costruzione insistenti nel garage n. {n:99} sito in {t:Rovigo} di proprietà dell'attore. Per l'effetto, il Sig. {a-l:Grossi} chiedeva che il Tribunale adito condannasse la {j-f:Prima} Spa a risarcire tutti i danni, patrimoniali e non, subiti dallo stesso attore e, in particolare, a titolo di danno patrimoniale, alla eliminazione dei vizi costruttivi da individuarsi e quantificarsi con apposita consulenza tecnica, alla somma di € {n:900,00} quale differenza tra i danni materiali risarciti e quelli richiesti in sede di mediazione e alla somma di € {n:3.000,00} a titolo di ulteriori danni patrimoniali ovvero da liquidarsi in via equitativa; a titolo di danno non patrimoniale il Sig. {a-l:Grossi} chiedeva che la {j-f:Prima} Spa venisse condannata alla somma di € {n:3.500,00} ovvero a quella diversa di liquidarsi in via equitativa, e dunque complessivamente alla somma di € {n:7.400,00} oltre alla eliminazione dei vizi costruttivi, ovvero alla somma che sarà ritenuta di giustizia.

Figure 4: Marked version of a Court proceeding.

u for alphanumeric strings (e.g. tax codes, abbreviations of provinces, licence plate number, etc.);

x for data that do not fall into any of the previous categories (they are re-placed with ###).

For each category of personal data identified by a tag, the program draws on a list of thousands of terms of the same category (male and female first names, surnames and toponyms[12]) in order to automatically replace them (the program chooses names beginning with the same initial as the original names in order to preserve euphony). Therefore, marking "Prima" as "juridical person" (j-f), as shown in Figure 4, instructs the program to remove this item and replace it with another legal person name from the list provided.

---

12 For the purposes of the project, it was not necessary to create subcategories of toponyms (i.e. distinguishing city names from country names).

Numeric and alphanumeric strings (such as licence plate number, fax and tele-phone numbers, etc.) are instead replaced with random numeric and alphanumeric strings of the same length. Moreover, since a consistency in temporal references within the text is crucial to use the documents also for legal studies (which always involve the reconstruction of the facts and the events of the trial), the substitution of dates follows a particular procedure: the program subtracts by default from the year a numerical value between a minimum of 5 and a maximum of 15 (the same throughout the analysis session), leaving instead month and day intact. How-ever, it is also possible to opt for a random substitution of dates.

The result of the pseudonymization process can be seen in Figure 5:

TRIBUNALE DI ROGOLO

R.G. N. 6183/2002

PER

La Perla Spa, in persona del legale rappresentante pro tempore, Sig. Maccabeo Renzullo, con sede in Rogolo, Via Jazelynn Tundis, 88, P.IVA 35068033487, rappresentata e difesa dall'Avv. Algeri Valeri (C.F. PTKLTZ49P50U275B, fax 4841/007983, pec ip6042@gmail.com) presso il cui Studio sito in Rogolo, Via Yilin Uggetti, 80, ha eletto domicilio giusta delega posta in calce al presente atto

CONTRO

Il Sig. Gardumo Zosimo con l'Avv. Adelaide Piccioni

COMPARSA DI COSTITUZIONE E RISPOSTA

Con atto di citazione notificato in data 15/10/2001, il Sig. Gardumo Zosimo conveniva in giudizio dinanzi l'intestato Tribunale la Perla Spa per sentire accertare e dichiarare la responsabilità della società convenuta per i vizi di costruzione insistenti nel garage n. 84 sito in Rogolo di proprietà dell'attore. Per l'effetto, il Sig. Gardumo chiedeva che il Tribunale adito condannasse la Perla Spa a risarcire tutti i danni, patrimoniali e non, subiti dallo stesso attore e, in particolare, a titolo di danno patrimoniale, alla eliminazione dei vizi costruttivi da individuarsi e quantificarsi con apposita consulenza tecnica, alla somma di € 387,67 quale differenza tra i danni materiali risarciti e quelli richiesti in sede di mediazione e alla somma di € 4.273,81 a titolo di ulteriori danni patrimoniali ovvero da liquidarsi in via equitativa; a titolo di danno non patrimoniale il Sig. Gardumo chiedeva che la Perla Spa venisse condannata alla somma di € 4.749,39 ovvero a quella diversa di liquidarsi in via equitativa, e dunque complessivamente alla somma di € 9.023,21 oltre alla eliminazione dei vizi costruttivi, ovvero alla somma che sarà ritenuta di giustizia.

Figure 5: Pseudonymised version of the document shown in Figure 4.

To keep consistency in substitutions within the text, all occurrences of the same term preceded by the same tag are replaced with the same fictional term throughout the document. Thus, in the example above, the program has replaced "Prima" with

"Perla" in all its occurrences within the proceeding.[13] This ensures conceptual-semantic consistency not only within a single document but, in the case of more than one document related to the same judgement, within all interrelated documents.

The readability of the text would certainly have been compromised if the document had been anonymised using traditional anonymising methods, as shown in Figure 6.

TRIBUNALE DI ▬▬

R.G. N. ▬/▬

PER

La ▬ **Spa, in persona del legale rappresentante** ▬, **Sig.** ▬ ▬, con sede in ▬, Via ▬, P.IVA ▬, rappresentata e difesa dall'Avv. ▬ (C.F. ▬, fax ▬, pec ▬) presso il cui Studio sito in ▬, Via ▬, ha eletto domicilio giusta delega posta in calce al presente atto

CONTRO

Il Sig. ▬ ▬ con l'Avv. ▬ ▬

COMPARSA DI COSTITUZIONE E RISPOSTA

Con atto di citazione notificato in data ▬, il Sig. ▬ conveniva in giudizio dinanzi l'intestato Tribunale la ▬ Spa per sentire accertare e dichiarare la responsabilità della società convenuta per i vizi di costruzione insistenti nel garage n. ▬ sito in ▬ di proprietà dell'attore. Per l'effetto, il Sig. ▬ chiedeva che il Tribunale adito condannasse la ▬ Spa a risarcire tutti i danni, patrimoniali e non, subiti dallo stesso attore e, in particolare, a titolo di danno patrimoniale, alla eliminazione dei vizi costruttivi e individuarsi e quantificarsi con apposita consulenza tecnica, alla somma di € ▬ quale differenza tra i danni materiali risarciti e quelli richiesti in sede di mediazione e alla somma di € ▬ a titolo di ulteriori danni patrimoniali ovvero da liquidarsi in via equitativa; a titolo di danno non patrimoniale il Sig. ▬ chiedeva che la ▬ Spa venisse condannata alla somma di € ▬ ovvero a quella diversa di liquidarsi in via equitativa, e dunque complessivamente alla somma di € ▬ oltre alla eliminazione dei vizi costruttivi, ovvero alla somma che sarà ritenuta di giustizia.

Figure 6: Anonymised version of the document shown in Figure 4.

Pseudonymization ensures a full protection of the confidentiality of natural and legal persons involved in the case, since the fictional data in the document do not make the persons or events mentioned recognisable. Moreover, the substitution legend is never made explicit by the program and is not exportable in any way by those who carry out the substitution. It is basically a "disposable" legend which is

---

13 For more examples cf. Fusco 2021: 33–34.

deleted at the end of the process and will never be retrievable afterwards (cf. Lombardi 2021).

Given the purpose of the project and the nature of the texts, it was not necessary to devise a de-pseudonymization procedure[14] which, by an inverse mechanism to pseudonymization, allows personal data uniquely associated with fictional data to be unambiguously recovered.

As mentioned above, text marking is functional for pseudonymization, but also for subsequent linguistic study. Therefore, other tags have been used that do not entail the substitution of the marked portion of the text, but they are preparatory to linguistic analysis. This is the case for ISO 639 codes, which are used preceded by f- (*foreign*), on a {f-eng: text} pattern, to mark loanwords from foreign languages or Latin. So, for example, "pro tempore", marked as Latinism in the text shown in Figure 4, is highlighted but not replaced in the pseudonymised text shown in Figure 5.

5 The data flow

Moreover, the described process is guided by a variable set of rules that can be configured according to the intended objectives: as we have already emphasised, in our case the processing of the texts is focused on both pseudonymization and linguistic analysis; procedures such as the forestierisms marking and the attention paid to the respect of phonosyntactic phenomena, even in the pseudonymization phase, contribute to the collection of metadata. As mentioned above, and further discussed below, other sources of metadata are the digital rich text format itself, which allows for the retrieval of typographical aspects, and other external tools, such as the POS taggers.

Thanks to this approach, which adds information only to take it away, the pseudonymization system becomes capable of reshaping the source document, with its purely typographical structure, into a semantically structured document. By making use of the various metadata sources included in the input, the system not only anonymises the document, but also converts it into a real TEI document. In this way, the complete process involves a decoding phase of the original format, a pseudonymization phase according to a variable set of rules, and finally the generation of a TEI document, accompanied by possible typographical renditions in HTML, to provide operators with immediate feedback on their work (see Figure 7).

---

14 Cf. Elger et al. 2010, Noumeir, Lemay and Lina 2007.

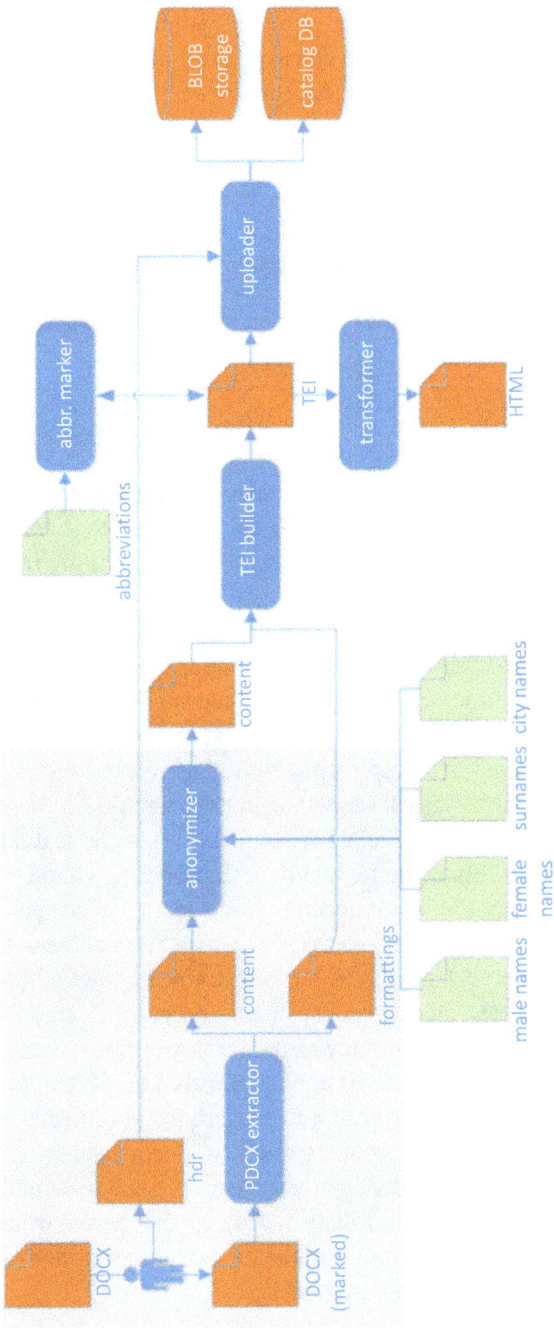

Figure 7: First part of the general data flow: the grey box delimits the protected area, from which no personal data may leave.

Indeed, document metadata come from multiple sources: first of all, the manual markup applied for the obscuring of personal information and for the annotation of useful aspects according to linguistic analysis; secondly, the word processing format (DOCX specifically) in which almost all documents are collected, and from which only a minimal subset of typographical information is extracted; and finally, the use of POS tagging systems, that makes it possible to obtain – with a good approximation – further metadata relating to the lemmatisation and morphological classification of each word. All this metadata must then find its way into the index that will feed the search engine base.

6 Search requirements

In order to meet a number of project requirements, a new open-source search engine (code-named Pythia[15]) has been introduced into the data flow that leads from Word documents to their pseudonymised TEI outputs. As describing this would require a paper in itself, here we will just provide an overview of its most useful features for this project.

First of all, the project requires a concordance-based approach, combined with a reusable index in the form of a database, which includes the documents themselves and their metadata. In this context, many of the widely-used text search engines are mainly focused on locating a document in a huge corpus with some relevance score, rather than on detailing the occurrences of each expression in its context.

Also, we need to integrate a virtually unlimited set of metadata in the textual search, which might even be larger than the textual data themselves, whatever their source, which is typically composite: input documents metadata, typographic attributes (e.g. italic, paragraph alignment, etc.) extracted from the original format, additional information provided by POS tagging, etc. All these metadata refer to different levels of analysis, have different models, and span over different regions of the text. Moreover, these regions are often not only nested, but even overlapping, thus going beyond the simple encoding capabilities of an XML-based document. Indeed, the indexing process must also act as a sort of hub, gathering and fitting data from completely different and independent sources into a uniform model: e.g. Excel files with document metadata; Word files that correspond to the texts and from which we can derive some formatting information; TEI tags that derive from manual tagging and extraction of metadata from documents; POS tags that derive from independent processing of plain text counterparts; etc.

---

15 See https://github.com/vedph/pythia for source code and documentation.

However, we still need a method of incorporating all this metadata into an index that provides a uniformly searchable surface, which can only be done in the context of a higher level of abstraction. It is precisely the higher level of abstraction provided by Pythia that allows not only words and their metadata, but also more extended text structures (such as sentences or – although this is not the case in this research – verses and stanzas),[16] with their metadata, to be handled in the same way. As already mentioned, the system has to take into account the obvious fact that such structures very often do not overlap at all, as they are on completely different levels of linguistic analysis: for instance, the syntactic tree of a text does not always correspond with its metrical organisation in verse or strophe, or with its cholometric arrangement at typographical level. Apart from this example, an essential requirement in this project is to be able to define certain essential structures that span several words, such as the sentence, because this allows contextual research in a more meaningful context, defined by syntax rather than a mechanical word count.

7 The Pythia search engine

The core idea of Pythia is a kind of de-materialisation of text (Fusi 2020). Traditionally, a full-text search index focuses on sequences of characters, variously extracted and filtered from a text, and possibly accompanied by additional metadata. The search focuses on matching such sequences, possibly limiting the results with the help of the metadata; in more powerful scenarios, more complex logic is added to handle contextual searches (e.g. searching for one word before another, or for two words in the same sentence, etc.).

In Pythia, however, although the basic mechanisms are the same, the focus shifts from this sequence of characters to an open set of objects, each with any number of metadata. In a sense, corpora, documents, sentences, words and other similar structures are all objects. Most of these objects include, among their metadata, a position with respect to the source document, but this is not a requirement (e.g. a document has no notion of position).

In this architecture, the search consists of finding all objects whose metadata match the specified criteria, within the specified subset of the index. Metadata are thus the endpoints through which users access objects. For example, searching for

---

16 Such examples are better understood in the context which first gave rise to the Pythia project, which is right a linguistic and metrical analysis system (Chiron), providing literally millions of highly specialized metadata to plain texts, ranging from whole lines (or sentences, in the case of rhythmic prose) to subphonematic traits.

the word "abbandonato (*left*)" means finding all occurrences of words whose attribute value (the attribute representing the text of a word) is equal to "abbandonato" (see Figure 8).

Figure 8: The Pythia search UI in the web client.[17]

Of course, an object corresponding to a word can have many other attributes, such as part of speech (detected by a POS tagger), number of syllables, number of letters, various linguistic classifications (e.g. foreign word, dialect word, technical term, personal name, or geographical name), etc. There is no limit to the attributes that we can assign to each element, and thus to the types of search we can perform.

Thus, an object such as a phrase is essentially no different from an object such as a word, and both can be searched in the same way. Moreover, a search not only includes the whole range of usual functions, such as logical operators, parentheses, wildcards, regular expressions, fuzzy matches and so on; it also provides a series of positional operators, which allow users to find one object within another, or partially overlapping, or at the beginning or end, etc.

In this context, a single word can be thought of as a point, and a longer structure as a segment; positional operators then allow users to find intersections, inclusions, etc. This makes it possible to combine all objects, such as words and phrases, into a unified search expression.

---

17 The query expression produces a number of results, shown in the typical KWIC grid. At the bottom, you can browse and read the document, eventually starting from any of the results (such as here, highlighted in yellow). The document map on the left is generated automatically from the document content. The portion of text displayed corresponds to the selected node in that map. Should you select the whole document (act node), the whole document will be shown.

This sort of de-materialisation of the text makes it possible to search an open and extremely varied set of metadata, regardless of the object to which they belong or their, often multiple, sources.

In this context, the indexing system must enable the same openness of the search. This is achieved by linking different types of reusable software modules into a pipeline, which manages the entire data processing flow from their source, whatever their location (file system, database, web resources, etc.) and their digital format (plain text, XML as TEI, database, etc.), to the final parser leading to index models, including filters, tokenizers, metadata extractors and calculators, text structure detectors, etc.

All these modules are often configurable and thus suitable for reuse; but they can also provide more strong logic in the combination of data sources. For instance, in the AttiChiari project, sentence boundaries are detected by any of the numerous filters in the indexing pipeline. Since the input format in this case is TEI, a generic sentence division algorithm (mainly based on punctuation) is combined with a configurable approach that also takes into account the nature of certain tags. Thus, a tag such as <head>, used for headings, is treated as a sentence even if there is no end punctuation. In this way, the system can use a sentence splitting module that exploits the information of both the text and its XML tags, if any. The same module can be reused whenever it is necessary to split sentences, both for text and XML documents, regardless of their dialect (as in this case TEI).

Other configurable components concern the mapping of the internal articulation of the text (e.g. division into sections, paragraphs, etc.). This provides functionalities such as an interactive navigation map of the text; the extraction of portions of text to be presented in a semantically congruent context (based on this same map); and the transformation of the original format of the text into a format intended for its presentation (typically HTML and CSS). In the case of the TEI texts discussed here, the transformation module simply uses an XSLT script which is provided to it as an operational parameter. These features thus provide a working tool in which users can switch from searching to reading without interruption.

In the context of this open indexing system, the peculiar nature of the treated texts led to a further evolution of the pseudonymiser to automatically identify abbreviations, which are too numerous for a reliance on manual tagging. In fact, the identification of sentence structures is essentially based on punctuation, although not exclusively, as in the case of documents with markup involving syntactic boundaries. Thus, the frequent presence of abbreviations containing full stops would constitute a major source of error. Therefore, in order to avoid potential problems in identifying sentences, and to provide a more comprehensible text for non-specialised users, the ability to automatically mark abbreviations was added to the pseudonymization system, which was then integrated into the indexing process.

This is, thus, a further example of how the very nature of this system is shaped by the needs of the analysis, which are mainly (but not exclusively) linguistic.

Finally, in the context of such an open and embeddable system, it should be added that its implementation is based on a set of standard and widely-used technologies: the index is nothing other than an RDBMS, which can be easily integrated and used, even bypassing the provided engine; the query language is defined by an ANTLR grammar; and the entire process leading from the original document in its format, to the model used for searching and reading, is configured in a modular pipeline, defined by an external file. The system can then be enriched simply by introducing new modules and chaining them into the pipeline, which can then be used to add new data sources from different analyses, merging them all into a uniform surface. In this way, the peculiar nature of the documents handled by this project and its requirements, both in linguistic and legal terms, can be well matched by the level of abstraction and the high modularity of the proposed solution.

8 Conclusions

The pseudonymization process adopted in the AttiChiari project, and illustrated in this contribution, combines the need to protect confidentiality with the need to have formally complete texts that can enable linguistic analysis and the identification of clear and effective forensic writing samples.

The approach discussed here serves, first and foremost, to ensure complete and non-reversible anonymisation of the data, which is not only an obvious legal requirement, but a fundamental aspect of gaining the trust of contributors to the corpus of documents. The fact that the software is only downloadable and usable locally by a restricted group of qualified operators, that the mapping of substitutions is temporary and unknown to the annotators themselves, and that the original documents are never placed on the network, is a further guarantee of the protection of privacy, as in this way the original texts are protected from any possible violation.

Secondly, the manual annotation of the operators, the additional information resulting from the conversion of the text format from DOCX to TEI (such as, for instance, typographical styles), and the additional information resulting from additional processes (such as the undoing of abbreviations or the third-party tagger), make it possible to offer the researcher or professional who needs to consult the database, perfectly readable, coherent and complete texts.

What now remains to be perfected is a search system capable of processing an open set of annotations, extended both on words and other linguistic structures (e.g. the sentence), and of providing a concordance search that integrates all these data sources on the same level, thus ending up with a set of tools whose practical and methodological value may transcend the boundaries of the individual research project.

References

Alpa, Guido. 2003. "Forensic Linguistics": il linguaggio dell'avvocato nell'evoluzione dell'ordinamento, dei metodi interpretativi, delle prassi e della tecnologia. In Alarico Mariani Marini (ed.), *La lingua, la legge, la professione forense*, 21–46. Milano: Giuffrè.

Candrilli, Fernanda. 2021. Il progetto di archiviazione e anonimizzazione. In Riccardo Gualdo & Laura Clemenzi (eds.), *Atti Chiari. Chiarezza e concisione nella scrittura forense*, 19–29. Viterbo: Sette Città.

Caponi, Remo. 2016. Il processo civile telematico tra scrittura e oralità. In Federigo Bambi (ed.), *Lingua e processo. Le parole del diritto di fronte al giudice*, Atti del Convegno (Firenze, 4 aprile 2014), 176–186. Firenze: Accademia della Crusca.

Cavallone, Bruno. 2012. Un idioma coriaceo: l'italiano del processo civile. In Barbara Pozzo e Federigo Bambi (eds.), *L'italiano giuridico che cambia*, Atti del Convegno (Firenze, 1ottobre 2010), 85–95. Firenze: Accademia della Crusca.

Clemenzi, Laura. 2021. L'interrogazione della base dati Atti Chiari. In Riccardo Gualdo & Laura Clemenzi (eds.), *Atti Chiari. Chiarezza e concisione nella scrittura forense*, 41–52. Viterbo: Sette Città.

Dalianis, Hercules. 2019. Pseudonymization of Swedish Electronic Patient Records Using a Rule-Based Approach. In Lars Ahrenberg & Beáta Megyesi, *Proceedings of the Workshop on NLP and Pseudonymization*, 16–23. Turku: Linköping Electronic Press. https://aclanthology.org/W19-6503.

Dell'Anna, Maria Vittoria. 2016. Fra attori e convenuti. Lingua dell'avvocato e lingua del giudice nel processo civile. In Federigo Bambi (ed.), *Lingua e processo. Le parole del diritto di fronte al giudice*, Atti del Convegno (Firenze, 4 aprile 2014), 83–101. Firenze: Accademia della Crusca.

Dell'Anna, Maria Vittoria (ed.). In press. *Lingua e scrittura forense. Storia, temi, prospettive*. Torino: Giappichelli.

Douglass, Margaret, Gari D. Clifford, Andrew Reisner, George B. Moody & Roger G. Mark. 2004. Computer-Assisted De-Identification of Free Text in the MIMIC II Database. *Computers in Cardiology* 31. 341–344. https://cinc.org/archives/2004/pdf/341.pdf.

Elger, Bernice S., Jimison Iavindrasana, Luigi Lo Iacono, Henning Müller, Nicolas Roduit, Paul Summers & Jessica Wright. 2010. Strategies for health data exchange for secondary, cross-institutional clinical research. *Computer Methods and Programs in Biomedicine* 99(3). 230–251. https://www.sciencedirect.com/science/article/pii/S0169260709003046?via%3Dihub.

Freddi, Maria. 2019. *Linguistica dei corpora*. Roma: Carocci.

Fusco, Francesca. 2021. Marcatura linguistica e tutela della riservatezza nello studio di un *corpus* di scritture forensi. In Riccardo Gualdo & Laura Clemenzi (eds.), *Atti Chiari. Chiarezza e concisione nella scrittura forense*, 29–40. Viterbo: Sette Città.

Fusi, Daniele. 2020. Text Searching Beyond the Text: a Case Study. *Rationes Rerum* 15. 199–230.

Fusi, Daniele. 2021. Digitalizzazione e marcatura XML degli atti. In Riccardo Gualdo & Laura Clemenzi (eds.), *Atti Chiari. Chiarezza e concisione nella scrittura forense*, 59–73. Viterbo: Sette Città.

Gualdo, Riccardo. 2011. Il linguaggio del diritto. In Riccardo Gualdo & Stefano Telve, *Linguaggi specialistici dell'italiano*, 411–477. Roma: Carocci.

Gualdo, Riccardo. 2021. Chiarezza e concisione negli atti processuali. In Riccardo Gualdo & Laura Clemenzi (eds.), *Atti Chiari. Chiarezza e concisione nella scrittura forense*, 11–18. Viterbo: Sette Città.

Gualdo, Riccardo & Laura Clemenzi (eds.). 2021. *Atti Chiari. Chiarezza e concisione nella scrittura forense*. Viterbo: Sette Città.

Gualdo, Riccardo & Maria Vittoria Dell'Anna. 2016. Per prove e per indizi (testuali). La prosa forense dell'avvocato e il linguaggio giuridico. In Giovanni Ruffino & Marina Castiglione (eds.), *La lingua variabile nei testi letterari, artistici e funzionali contemporanei. Analisi, interpretazione, traduzione*, Atti del XIII Congresso SILFI (Palermo, 22–24 settembre 2014), 623–635. Firenze: Cesati.

Lombardi, Giulia. 2021. I vantaggi del programma *an-tool*. In Riccardo Gualdo & Laura Clemenzi (eds.), *Atti Chiari. Chiarezza e concisione nella scrittura forense*, 29–40. Viterbo: Sette Città.

Mariani Marini, Alarico (ed.). 2003. *La lingua, la legge, la professione forense*. Milano: Giuffrè.

Mortara Garavelli, Bice. 2001. *Le parole e la giustizia. Divagazioni grammaticali e retoriche su testi giuridici italiani*. Torino: Einaudi.

Mortara Garavelli, Bice. 2003a. L'oratoria forense: tradizione e regole. In Alarico Mariani Marini & Maurizio Paganelli (eds.), *L'avvocato e il processo. Le tecniche della difesa*, 66–91. Milano: Giuffrè.

Mortara Garavelli, Bice. 2003b. Strutture testuali e stereotipi nel linguaggio forense. In Alarico Mariani Marini (ed.), *La lingua, la legge, la professione forense*, 3–19. Milano: Giuffrè.

Noumeir, Rita, Alain Lemay & Jean-Marc Lina. 2007. Pseudonymization of Radiology Data for Research Purposes. *Journal of Digital Imaging* 20(3). 284–295. https://link.springer.com/article/10.1007/s10278-006-1051-4.

Oksanen, Arttu, Minna Tamper, Jouni Tuominen, Aki Hietanen & Eero Hyvönen. 2019. ANOPPI: A Pseudonymization Service for Finnish Court Documents. In Michał Araszkiewicz & Víctor Rodríguez-Doncel (eds.), *Legal Knowledge and Information Systems*. JURIX 2019: The Thirty-second Annual Conference, 251–254. Amsterdam: IOS Press. https://ebooks.iospress.nl/publication/53682.

Rodríguez-Puente Paula, Cristina Blanco-García & Iván Tamaredo. 2019. Mark-up and Annotation in the Corpus of Historical English Law Reports (CHELAR): Potential for Historical Genre Analysis. *Atlantis. Journal of the Spanish Association for Anglo-American Studies*. 41(2). 63–84. https://www.atlantisjournal.org/index.php/atlantis/article/view/617.

Sabatini, Francesco. 2003. Dalla lingua comune al linguaggio del legislatore e dell'avvocato. In Alarico Mariani Marini & Maurizio Paganelli (eds.), *L'avvocato e il processo. Le tecniche della difesa*, 3–14. Milano: Giuffrè.

Santulli, Francesca. 2008. La sentenza come genere testuale: narrazione, argomentazione, performatività. In Giuliana Garzone & Francesca Santulli (eds.), *Il linguaggio giuridico. Prospettive interdisciplinari*, 207–238. Milano: Giuffrè.

Visconti, Jacqueline. 2018. La chiarezza tra superfluo e necessario. In AA.VV., *Breviario per una buona scrittura*, 15–19. Roma: Ministero della Giustizia. https://www.federnotizie.it/wp-content/uploads/2018/10/BREVIARIO_ATTI_PROCESSUALI.pdf.

Visconti, Jacqueline. 2021. Introduzione. In Riccardo Gualdo & Laura Clemenzi (eds.), *Atti Chiari. Chiarezza e concisione nella scrittura forense*, 9–10. Viterbo: Sette Città.

Visconti, Jacqueline. 2022. *Studi su testi giuridici. Norme, sentenze, traduzione*. Firenze: Accademia della Crusca.

Alessandra Lombardi & Karin Luttermann

Legal specialists telling their stories on the internet: A comparative analysis of professional self-portrayals on German and Italian law firm websites

One might expect that linguistics – the scientific study of language – would have much to offer the legal profession, which is so often occupied with the interpretation of legal texts. (Peter M. Tiersma 1995: 1095)

1 Introduction

Linguistics is concerned with language, but also with communication: broader than language, communication also includes paraverbal and non-verbal features. We do much of our communication through language, choosing our words and the way we use them; but we also communicate with gestures and facial expressions, as well as with our overall appearance and, possibly, with images or other visual support. People in everyday life, including in business and professional settings, can only come together through communication: it is the fundamental basis of any relationship, including that of a company with its employees and clients. This applies not only internally, with regard to coordination of work procedures and principles of leadership, but also externally, with a view to a positive presentation of the company and its products, as well the acquisition of clients (see Keller 2009: 19–20). Without well-planned and effective communication, companies cannot do business successfully and set themselves apart from their competitors. That is why it is not only what is said that counts, but also how it is said and through what medium.

This study deals with staff presentations through digital media. Our specific focus is on so-called 'boutique' law firms. These are comparatively small, highly-specialised law firms, filling a niche and aiming to ensure a high level of quality in the legal advice which they provide for their clients. The specialisation of a boutique law firm can be seen in relation to the areas of law it deals with, and/or its clients. As is the case with a specialist shop, the clients know that they are in the right place for a particular legal problem (see Rossbach 2009: 69–76).

Ours is a comparative study of epistemic cultures (according to Knorr-Cetina 1999 and Liebert 2016, this means examining the construction of special knowledge in a discipline, on the basis of a common conceptual and value-oriented background).

https://doi.org/10.1515/9783111048789-004

Specifically, the focus is on practices of self-presentation by legal specialists on German and Italian law firm websites. The aim is to make a pragmatic analysis of how texts are used to build up a professional image in the two legal cultures. The investigation is guided primarily by the following questions:

1. Which communication formats (conventional *curriculum vitae/CV, biographies/ bios*) do lawyers use in the two cultures for their professional self-profiling? Where are common features to be found, and where are the differences?

2. What influence do websites, as multimodal platforms of communication, have on the recruitment and marketing of legal specialists advertising their expertise and services on the internet?

3. What kind of professional stereotypes (*auto-, hetero-, meta-stereotypes* – see Thiele 2015: 30) occur, as a reflection of specific cultural standards (cultural influence, conceptualisations of the legal profession)?

These questions are examined as follows: from text pragmatics (Section 2) and multimodality (Section 3), we move on through description of the corpus design and method (Section 4) to the quantitative results of our pilot study (Sections 5 and 6). Concluding remarks can be found in Section 7.

2 Text pragmatics

Text linguistics is a constantly evolving discipline of modern linguistics, which addresses the specificities of new media in the more specific setting of hypertext linguistics (see Section 3). It is considered a powerful field of research, in terms not only of theory and methodology but also of interdisciplinarity and practical applications (see Fix 2009: 11). Its main subjects of study are texts and text types. Within the framework of the communicative-pragmatic approach to texts, the description of text types is fundamental. This description is a central feature of text pragmatics, identifying as it does the connections between functionally-situationally determined communicative actions and their underlying formulation patterns (see Gansel and Jürgens 2009: 53).

The analysis of text types is concerned with patterns of action, as they have developed language- and culture-specifically, for recurring communicative purposes. A single text is assigned to a text type on the basis of its communicative function. Rolf (1993) and Brinker (2005) have classified different text functions, based on Searle's speech act theory. With regard to the self-presentation of companies and law firms, *job advertisements (job ads)* are of particular interest here. Our contribution builds on earlier comparative studies examining the text-pragmatic features of legal job advertisements (see Luttermann 2017, 2018; Luttermann and Engberg 2017; Lombardi, in preparation). The *job ad* text type displays strongly homogeneous communicative functions across cultures (primarily the appellative function, as a call for applications to fill a vacant position).

In addition, advertisements then develop into instruments of employer branding (see Nielsen, Luttermann and Lévy-Tödter 2017). This development makes it particularly worthwhile to investigate the culture-specific dimensions – not only with regard to common professional values and attitudes in different national contexts, but also in relation to specific accents of local colour (see articles in Schieblon 2009). Apart from retention (keeping staff) and development (the company as such), employer branding also includes attraction (recruitment of staff). Self-presentation (self-promotion) of the company is the means for recruiting qualified staff. The category of attraction is also significant for the *curriculum vitae* text type, as lawyers talk about their career with the intention of acquiring new clients. They want to persuade addressees to call on their services, rather than those of competitors (see Section 3).

3 Multimodality

Text linguistic research sees the CV as a text type characterised by a high degree of standardisation in relation to macro- and micro-structure and fulfilment of communication function – these features being common to different languages and cultures (see Zhao 2011). The analysis of legal job ads embedded in law firm websites suggests that the World Wide Web, as the "hypertext platform of the Internet" (Storrer 2008: 317), has considerable effects on the way legal experts present their professional services. Media are in a continuous state of interplay (see Holly 2011: 158). Through remediation, content is constantly linked and made available in new and different ways, and new text-type variants are produced – for example, by deliberately deviating from conventions in order to attract the addressee's attention.

The traditional CV (whether in discursive or in table form) is changing. As far as we can see, there are, as yet, no studies of what multimodal means CVs use in order to change the conventionalised format and adapt it to the digital medium. Drawing on this digital dimension, we use a broad concept of text, regarding it as a "multimodal semiotic unit" (Bendel Larcher 2015: 50). Text analysis of the CV thus comprises not only linguistic signs, but also paraverbal and non-verbal signs and multimodal elements (such as written and spoken language, still and animated images, as well as various types of audio material). From our point of view, a multimodal understanding of text is needed for the identification of text-type variants and semiotic access to texts (Sections 4 and 5).

In linguistic pragmatics, meaning is not understood as an inherent property of signs, but as the product of an active process of binding a certain content to a sign form by sign users. They possess certain knowledge for the negotiation of meaning. This knowledge is incorporated into texts (see Klug 2016: 165–166; Luttermann 2010: 150–151). Against this background, the central point of reference for our analysis is the multimodal nature of lawyer biographies on websites. A law firm has to provide information as to why, of all law firms, it is this particular one that should be mandated, rather than its competitors. How the unique selling proposition (USP) is communicated to the client, and which potential advantages of the text sort are still untapped, is an area of research that urgently needs to be addressed (see Section 1).

4 Pilot study: Online self-portrayals of German and Italian legal specialists

In order to gain an initial insight into the way legal professionals introduce themselves on the web, we designed a small-scale exploratory study with the aim of comparing German and Italian self-presentations on law firm websites. Figure 1 shows an example of a legal self-portrayal embedded in the website of a German boutique law firm, with the prototypical structure displaying a half-length photograph and factual information about the legal specialist's profile.

As outlined in the introduction, we focused on boutique law firms as a flexible and agile legal business model which is gradually expanding, both in Germany and in Italy. Being small-sized enterprises operating in a very competitive arena, boutique law firms need unique brand positioning, and a clear and effective digital strategy may be a key driver in this process.[1]

---

1 It is interesting to note that at the time of writing (May 2022), the website design of the boutique law firm from which Figure 1 was taken in December 2021 has been completely revised (see https://www.ahlers-vogel.de/anwaelte-notare/dr-birgit-berninghausen/, last access 12.05.2022). This confirms a dynamic management attitude, aimed at exploiting the potential of digitalisation as a fundamental part of corporate branding strategy.

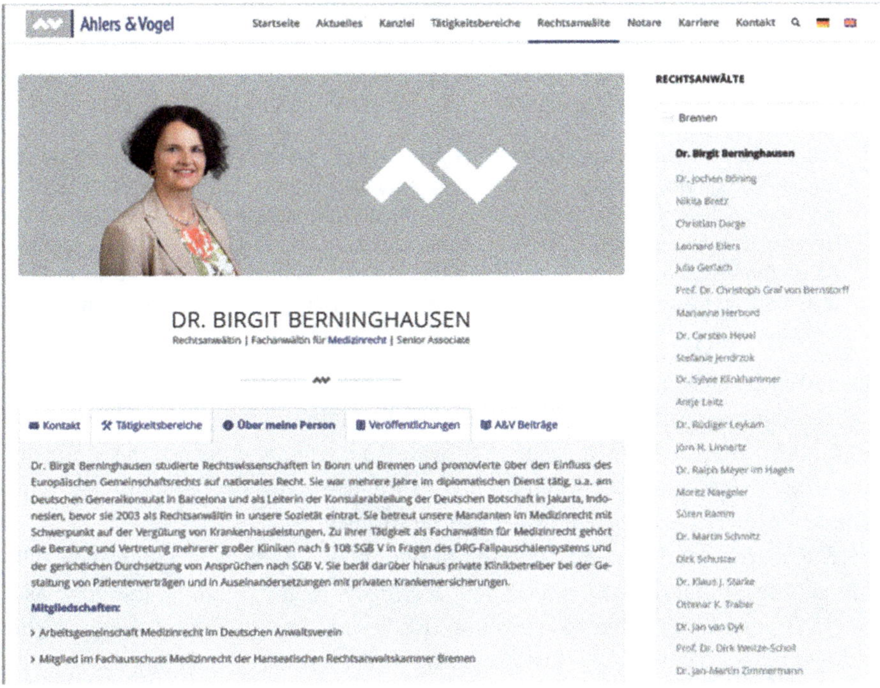

Figure 1: Example of self-portrayal of a German lawyer (https://www.ahlers-vogel.de, last access 10.12.2021).

4.1 Corpus design

A literature review, performed before setting up the corpus, revealed that the terminology used in Italian and German to designate boutique law firms is not yet consistent and may still be evolving. This lack of uniformity is reflected in the coexistence of several synonyms in contemporary usage (e.g., *Boutique – Boutique-Kanzlei – Boutique Kanzlei – Kanzleiboutique* in German; *studio legale boutique – studio-boutique – boutique legale – boutique del diritto – boutique giuridica* in Italian). The dynamic nature of the naming process and the related terminological uncertainties (see Temmerman and Campenhoudt 2014: 1–13) are also reflected in the different definitional approaches used in the current literature on this topic in Germany and Italy. German definitions hint at a more "management-oriented" perspective, while Italian definitions suggest a more "marketing-oriented" one, as illustrated in the following:

Dies sind spezialisierte Kanzleien, die auf ein oder mehrere Rechtsgebiete bzw. auf bestimmte Mandanten fokussiert sind. Boutiquen agieren in der Regel überregional und arbeiten für mittelständische und große Wirtschaftsunternehmen. Oftmals entstanden sie durch Ausgründungen, sogenannte 'Spin-Offs', von Großkanzleien. (Schieblon 2019: 4)

Fino ad arrivare agli studi legali di nicchia, quelli che per precisa volontà hanno scelto di occuparsi solo di una determinata materia presentandosi sul mercato come super esperti di quell'ambito del diritto. Molto spesso sono *i cosiddetti studi boutique* [emphasis ours]. [. . .] A che tipo di ristorante assomigliano queste realtà? Azzardiamo l'accostamento con una pizzeria gourmet, dove uno dei cibi più conosciuti e consumati in tutto il mondo diventa qualcosa di diverso per presentazione, ingredienti e prezzo! (https://www.4clegal.com/marketing-legale/mondo-studi-apparentemente-uguali-qual-vera-identita, last access 10.12.2021)

This terminological variation suggests that the concept itself is still quite fluid and context-dependent, making identification of suitable sources for corpus candidates a rather challenging task. In order to provide an adequate empirical basis for our pilot study, we adopted a bottom-up procedure, using the above-mentioned synonyms as keywords to inform our web-search strategy.

The search results mainly pointed to two different kinds of potentially relevant sources, namely *legal recruitment platforms* (using *boutique law firm* as a kind of Unique Employment Proposition (UEP) – i.e., as a label to distinguish this business model from others, thus providing professionals with a compass to navigate the legal job market) and *law firm websites* (using the term *boutique law firm* as a kind of USP – i.e., as a paramount part of a broader marketing strategy, highlighting the advertiser's core expertise and distinctive identity in the legal services sector).

It can be seen that, as early as the sample identification phase of our study, one potential area of investigation stands out as particularly worthy of attention: whether the selected firms distinguish between UEPs and USPs (see Stumpf 2017: 82); and, above all, to what extent staff presentations on their websites feed into these distinctive positioning strategies (see Section 6).

In order to collect a suitably balanced and comparable data set,[2] we set out to search both sources. This enabled us to build an ad hoc, exploratory corpus, including 100 legal professional profiles from 50 German and 50 Italian boutique law firm websites. The texts in our sample were gathered from the website sections devoted

---

2 Due to the exploratory nature of our study, our sampling strategy was mainly guided by the following two "external criteria": the domain (content type, i.e. the subject field) and the medium (the context of publication, the types of text and their main communicative functions). Other variables, such as the size or area of specialisation of the law firm, or the age and gender of the professionals, were not taken into account for the time being, but will be considered at a later stage when the composition of the corpus is refined in view of the full-scale project.

to the presentation of the professional staff offering legal services in the selected organisations.[3]

4.2 Multimodality and digitalisation as key components of the communicative strategy

A preliminary analysis of our sample highlighted the inherent relevance of multimodality as a central feature of online legal self-presentations, suggesting potential research questions to be addressed within a full-scale study. Detailed multimodal analyses (see Diekmannshenke, Klemm and Stöckl 2011: 9) are needed to investigate such points as the following:

a. account for the pragmatic effects of the interplay between text and image (e.g., the combination of lawyers' profile pictures and quotes from the lyrics of their favourite songs[4]);

b. explore the different communicative functions performed by digital animation and audiovisual content (e.g., the application of animation effects to miniature portraits of the staff members, conveying a "team-friendly attitude" as they look around and turn a benevolent, smiling gaze towards whichever colleague the cursor is moved to;[5] or the use of zoom-in effects, with the lawyer emerging from a blurred background and moving into focus, suggesting a "customer-friendly attitude"[6]);

c. assess the impact of specific interactive elements turning the users into active participants in the communication process (e.g., the use of dynamic displays, letting users customise their access to the information provided on the legal specialists, according to the desired degree of detail[7]).

---

3 The website sections that we examined in the search for German and Italian lawyers' profiles were entitled as follows: *Wir sind / Experten / Rechtsanwälte / Anwälte / Über uns / (Unser) Team and (Gli) Avvocati / (I) Professionisti / Team / Persone / Associati & partner.*
4 https://www.lcalex.it/persone/alessia-ajelli/ (last access 10.12.2021).
5 https://www.kmlz.de/de/team (last access 10.12.2021).
6 https://www.glademichelwirtz.com/team/dr-felix-bangel/ (last access 10.12.2021).
7 https://lambrecht.eu (last access 10.12.2021).

5 Comparing German and Italian legal self-presentations: some preliminary results

The empirical analysis carried out on both the German and Italian subcorpora allowed us to identify *three main categories* of legal self-portrayals:

1. the *CV-like presentation*, having the prototypical structure of a plain text organised in thematic sections, with boldfaced titles displaying the lawyers' personal data as well as their most salient academic and professional credentials;
2. the *hybrid* type, comprising features of the CV format and a descriptive profile in the third person, focusing on the lawyer's specific areas of expertise;
3. the *professional biography*, as a first-person narrative reconstructing (not necessarily in chronological order) the most significant stages in the lawyer's career and emphasising his/her professional values and distinguishing qualifications.

The corpus also displays a few examples of the integration of types 1 and 3 within the same self-presentation. Combining different modes of communication in this way is particularly interesting, as it offers two distinct but complementary perspectives on the lawyer's profile: a more objective and formal one through the CV, and a more personal and informal one through the biosketch. These different views of the same object are visually reinforced by a specifically tailored digital design, as well as by the posture in which the lawyer is represented in the two photos (see Figure 2).

The juxtaposition of the two modes of self-presentation reveals a clear visual strategy that is reflected in the texts alongside. The photograph accompanying the biography features touches of vivid colour, and shows the lawyer in a relaxed moment during his daily work routine, in a homelike working space. This visual representation communicates a professional, yet relatively informal and spontaneous attitude, enhancing the picture's immediacy and the impression of its subject's social closeness to the recipient. The same tendencies are consistently mirrored in the engaging, value-oriented and rather confidential tone of the text in which the lawyer tells his professional story – i.e., describing his team as a "second family". The communicative impact of the CV-like mode of presentation is completely different. The half-length photograph in suit and tie, as well as the sober, muted colours, clearly support the intention of conveying a more serious and formal image of the lawyer – in other words, his "business side", as reflected in the CV's objective statement of his professional experience and legal skills.

A quantitative review, aimed at assessing which of the three categories of legal self-presentation is more frequent in our corpus, provides a tentative basis for identifying the greater frequency of the *hybrid* text type: this can be seen in

Figure 2: Example of biography and CV of an Italian lawyer (https://lipanicatricala.it/avvocati/roberto-ferraresi/, last access 10.12.2021).

both the German and Italian samples (7 CV-like presentations, 39 *hybrids* and 4 bios in the German subcorpus; 9 CV-like presentations, 28 *hybrids* and 13 bios in the Italian subcorpus).

The preliminary content analysis of the selected websites showed a rather homogeneous communicative approach, with several cross-cultural features, but also a touch of local colour. The most recurrent topics, which can possibly be interpreted as a reflection of changing emphases in the legal profession's mode of communication *across cultures*, are clearly linked to the branding strategy of boutique law firms offering qualified, exclusive, customised legal services. These common features include:

a.   relationship-oriented communication and a high level of customer care;
b.   a positive, dynamic and light-hearted professional and interpersonal working atmosphere;
c.   a special focus on the work-life balance, and a cooperative team spirit;
d.   the creation of modern, functional workspaces (with some Italian firms showing great sensitivity to beauty and aesthetics, arranging their workspaces as fine *art galleries*[8]),
e.   the benefits of flexible, customer-tailored information services (especially by German firms, which often complement the lawyers' profiles with links to their posts in the in-house corporate blog or magazine, explaining specific legal matters to a non-specialist audience while emphasising the author's subject-field expertise[9]).

A touch of *Italian local colour* can be noticed in the special focus on interpersonal values and social skills (friendliness, approachability, sociability), probably aimed at bridging the distance between the expert giving legal advice and the typical Italian client: the latter, often intimidated by legal complications, will almost certainly prioritise not only competent and focused professional legal support, but also a close personal relationship based on trust and reliability.[10] *German local colour* can be related to the particular emphasis placed on expertise (on references, rankings and reputation), which is consistent with the findings of previous studies on German legal job ads (see Luttermann 2017; Luttermann and Engberg 2017). In order to communicate high professional standards, German lawyers often list awards and ratings by authoritative reviewers, acknowledging their professional skills as top-level specialists.[11] The aim is clearly twofold: to gain the trust of potential clients and to build brand credibility.

According to a preliminary analysis of multimodality in our sample, the German firms show a clear preference for a rather conventionalised self-presentation form (the *hybrid* text type), and seem to make smarter and more ample use of digital multimedia tools in order to achieve their communicative goals. In German self-presentations, the interactive elements are not merely visible as icons pointing to the relevant profile or page on social media sites, but are often integrated into the

---

8 See https://ricerca.repubblica.it/repubblica/archivio/repubblica/2016/11/09/lo-studio-legale-diventa-galleria-dartePalermo08.html (last access 10.12.2021).

9 Delivering complex legal knowledge in a clear, comprehensible and client-friendly way is conveyed as a fundamental part of the customer service offered – see, for example, https://zametzer-law.de/ (last access 10.12.2021).

10 See, for example, https://www.alfierizullino.it (last access 10.12.2021).

11 See, for example, https://www.lutzabel.com/anwaelte/dr-wolfgang-abel (last access 10.12.2021).

presentation text itself – e.g., in the form of an explicit invitation to access online platforms with user reviews of the law firm's services.[12] Moreover, video portraits on German websites are a valuable example of remediation producing specific communicative effects (see Section 3). One example of this is the emergence of a more informal and personal tone (in comparison with the rather impersonal and sober style of the textual self-presentations they are often associated with); another is the inclusion of new content items for positive stereotyping – e.g., *auto-stereotypes* (when lawyers talk about their professional self-image and experience), or *hetero-* and *meta-stereotypes* (when they refer to laypersons' perception of the legal profession).[13]

The content analysis of the Italian subcorpus shows more limited use of digital and multimedia tools, with greater preference for biography as a self-presentation form. Biographies on Italian boutique law firm websites are mainly characterised by highly person-centered, partly unconventional narratives, often blurring the boundaries between professional and private lives. In the biography text, the lawyer usually portrays himself/herself as a *highly skilled* specialist and a *value-driven* person, thus underscoring the relevance of personal ethics to professional ethics.[14]

6 Exploiting the communicative potential of legal self-portrayals to improve the brand identity of boutique law firms

The findings of our pilot study suggest that legal self-portrayals are not simply an emerging digital self-presentation form mirroring the changing attitudes towards the professional and social role of specialists in the new legal landscape: they may also be seen as effective communicative tools, enabling law firms to shape and develop their brand identity and strategy in today's internet-driven market.

Being a particular kind of *professional storytelling*, biographies can serve different pragmatic and rhetoric functions. For example, they can be a key component of a customer-driven marketing strategy (as when legal specialists communicate immediacy in describing their new ideal lawyer, challenging traditional professional stereotypes and placing human and interpersonal qualities at the heart of

---

12 See, for example, https://www.bronhofer.de/anwaelte/bronhofer (last access 10.12.2021).
13 See, for example, https://cms.law/de/deu/people/gerlind-wisskirchen (last access 10.12.2021).
14 See, for example https://lipanicatricala.it/avvocati/francesca-sbrana and https://kanzlei.wien bergs.com/ueber-mich.html (last access 10.12.2021).

their professional practice[15]). Alternatively, they can also be part of an employer branding strategy – for example, by the lawyer's self-description of their prior employment history: this provides a vehicle for presentation of the law firm's core values and professional standards, thus addressing and attracting potential candidates who may fit into this value framework[16]). A detailed analysis of a larger corpus could help assess whether this combination of strategies is a characteristic feature of legal self-portrayals in the category examined.

Further investigation is certainly needed to address issues of text typology, but our preliminary results already indicate that:

a. online self-presentations play a significant role in conveying the branding strategies of boutique law firms trying to gain a competitive edge in today's legal market;

b. detailed cross-cultural analysis of online legal self-portrayals can help to identify not only the professional standards typical of a given legal culture, but also the specific impact of the digital medium on the merging of legal communication styles across languages and cultures.

7 Concluding remarks

Text types are culturally shaped. This cultural influence is essentially based on its historical development within a language community. In a semiotic sense, text types function not merely as conventional linguistic patterns, but rather as patterns of communication. Turning to a multimodal understanding of text (instead of a mere linguistic comparison) makes it possible to investigate text types holistically, according to their verbal, paraverbal, non-verbal and extra-verbal features. Our analysis of lawyers' self-portrayals, in German and Italian boutique law firms, has shown how they have evolved into a new digital subgenre of self-presentation, affording a multifunctional complement to more traditional formats such as the written CV.

In the next stages of our research, the self-presentations of lawyers will be analysed in relation to specific variables (comprehensibility and impact on target groups, significance for knowledge transfer), by examining the response they elicit from a sample of internet users. The results of the analysis will form the

---

15 See, for example https://lipanicatricala.it/avvocati/damiano-lipani/ and https://www.anwalt-daum.de/ zur-person/ (last access 10.12.2021).

16 https://steueranwalt.de/anwaelte/dr-klaus-olbing (see, in particular, the section entitled "Mein Weg zu Streck Mack Schwedhelm") (last access 10.12.2021).

basis for practical recommendations to boutique law firms interested in leveraging their online presentations to best effect. We would like to stimulate an interdisciplinary dialogue between science and practice, in order to optimise digital communication between lawyer and client as an important contributory factor to the enhancement of trust in the legal profession.

References

Bendel Larcher, Sylvia. 2015. *Linguistische Diskursanalyse. Ein Lehr- und Handbuch.* Tübingen: Gunter Narr.

Brinker, Klaus. 2005. *Linguistische Textanalyse. Eine Einführung in Grundbegriffe und Methoden.* 6. Auflage. Berlin: Erich Schmidt.

Diekmannshenke, Hajo, Michael Klemm & Hartmut Stöckl (eds.). 2011. *Bildlinguistik.* Berlin: Erich Schmidt.

Fix, Ulla. 2009. Stand und Entwicklungstendenzen der Textlinguistik (I). In: *Deutsch als Fremdsprache* 46, 11–20.

Gansel, Christina & Frank Jürgens. 2009. *Textlinguistik und Textgrammatik. Eine Einführung.* 3. Auflage. Göttingen: Vandenhoeck & Ruprecht.

Holly, Werner. 2011. Medien, Kommunikationsformen, Textsortenfamilien. In Stephan Habscheid (ed.), *Textsorten, Handlungsmuster, Oberflächen. Linguistische Typologien der Kommunikation,* 144–163. Berlin: De Gruyter Mouton.

Keller, Rudi. 2009. Die Sprache der Geschäftsberichte: Was das Kommunikationsverhalten eines Unternehmens über dessen Geist aussagt. In Christoph Moss (ed.), *Die Sprache der Wirtschaft,* 19–44. Wiesbaden: Springer.

Klug, Nina-Maria. 2016. Multimodale Text- und Diskurssemantik. In Nina-Maria Klug & Hartmut Stöckl (eds.), *Handbuch Sprache im multimodalen Kontext,* 165–189. Berlin: De Gruyter Mouton.

Knorr-Cetina, Karin. 1999. *Epistemic cultures: How the sciences make knowledge.* Cambridge MA: Harvard University Press.

Liebert, Wolf-Andreas. 2016. Wissenskulturen. In Ludwig Jäger, Werner Holly, Peter Krapp & Samuel Weber (eds.), *Sprache – Kultur – Kommunikation / Language – Culture – Communication. Ein internationales Handbuch zu Linguistik als Kulturwissenschaft / An international handbook of linguistics as a cultural discipline,* 578–586. Berlin: De Gruyter Mouton.

Lombardi, Alessandra. (In preparation). Stellenanzeigen im Sprach- und Kulturkontrast. Zur fachkommunikativen Komplexität juristischer Stellenangebote als kulturbedingte Handlungspraktiken. In Albert Busch & Karin Luttermann (eds.), *Professionskommunikation als Sprache im Beruf.* Hildesheim: Universitätsverlag Hildesheim.

Luttermann, Karin. 2010. Verständliche Semantik in schriftlichen Kommunikationsformen. In *Fachsprache* 32 (3–4), 145–162.

Luttermann, Karin. 2017. Unsere Stärke: Sie! – Zielgruppenansprache und Textbausteine in Stellenanzeigen als ein Konzept von Werbung. In Martin Nielsen, Karin Luttermann & Magdalène Lévy-Tödter (eds.), *Stellenanzeigen als Instrument des Employer Branding in Europa. Interdisziplinäre und kontrastive Perspektiven,* 55–80. Wiesbaden: Springer VS.

Luttermann, Karin. 2018. Kommunikativ-funktionale Analyse von werbenden Gebrauchstexten in der Wirtschaft. In Kerstin Kazzazi, Karin Luttermann, Sabine Wahl & Thomas Fritz (eds.), *Worte über Wörter. FS Elke Ronneberger-Sibold*, 301–318. Tübingen: Stauffenburg Verlag.

Luttermann, Karin & Jan Engberg. 2017. Kulturkontrastive deutsch-dänische Textanalyse von sprachlichen Handlungen in juristischen Stellenanzeigen. In Martin Nielsen, Karin Luttermann & Magdalène Lévy-Tödter (eds.), *Stellenanzeigen als Instrument des Employer Branding in Europa. Interdisziplinäre und kontrastive Perspektiven*, 107–131. Wiesbaden: Springer VS.

Nielsen, Martin, Karin Luttermann & Magdalène Lévy-Tödter (eds.). 2017. *Stellenanzeigen als Instrument des Employer Branding in Europa. Interdisziplinäre und kontrastive Perspektiven*. Wiesbaden: Springer VS.

Rolf, Eckard. 1993. *Die Funktion der Gebrauchstextsorten*. Berlin: De Gruyter Mouton.

Rossbach, Oliver. 2009: Boutique-Kanzleien: Mit Spezialisierung zum Erfolg. In Claudia Schieblon (ed.), *Kanzleimanagement in der Praxis. Einführung und Management für Kanzleien und Wirtschaftsprüfer*, 69–82. 4. Auflage. Wiesbaden: Springer Gabler.

Schieblon, Claudia (ed.). 2019. *Kanzleimanagement in der Praxis. Einführung und Management für Kanzleien und Wirtschaftsprüfer*. 4. Auflage. Wiesbaden: Springer Gabler.

Schieblon, Claudia. 2019. Management in Kanzleien. In Claudia Schieblon (ed.) *Kanzleimanagement in der Praxis. Einführung und Management für Kanzleien und Wirtschaftsprüfer*, 1–12. 4. Auflage. Wiesbaden: Springer Gabler.

Storrer, Angelika. 2008. Hypertextlinguistik. In: Nina Janich (Ed.): *Textlinguistik. 15 Einführungen*, 315–331. Tübingen: Narr.

Stumpf, Marcus. 2017. Employer Branding versus Consumer Branding – (Stellen-)Anzeigen im Vergleich. In Martin Nielsen, Karin Luttermann & Magdalène Lévy-Tödter (eds.), *Stellenanzeigen als Instrument des Employer Branding in Europa. Interdisziplinäre und kontrastive Perspektiven*, 81–103. Wiesbaden: Springer VS.

Temmerman, Rita & Marc van Campenhoudt. 2014. Introduction: Dynamics and terminology. An interdisciplinary perspective on monolingual and multilingual culture-bound communication. In Rita Temmerman & Marc van Campenhoudt (eds.), *Dynamics and Terminology*, 1–14. Amsterdam: Benjamins.

Thiele, Martina. 2015. *Medien und Stereotype. Konturen eines Forschungsfeldes*. Bielefeld: transcript.

Tiersma, Peter M. 1995. The ambiguity of interpretation: Distinguishing interpretation from construction. In *Washington University Law Quarterly* 73 (3), 1095–1101.

Zhao, Jin. 2011. Kulturspezifik, Inter- und Transkulturalität von Textsorten. In Stephan Habscheid (ed.), Textsorten, Handlungsmuster, Oberflächen. Linguistische Typologien der Kommunikation, 123–143. Berlin: De Gruyter Mouton.

Giuliana Diani
Stance features in legal blogging

1 Introduction

It is well documented that digital technologies have brought profound changes to scholarly communication in recent years. As Anesa and Kulbicki (2022: 5) point out, law as a discipline and practice represents no exception to this innovation and the communication of law has been affected by it. Among the most significant of these changes has been the emergence of blogging as a vehicle for knowledge transfer to the wider community of the Internet, with both expert and non-expert readers. Within the legal field, "blogs have begun to affect the delivery of legal education, the production and dissemination of legal scholarship, and the practice of law" (Caron 2006: 1033). Berman (2006: 16), for example, found that "blogs can become an accepted medium for law professors to develop and disseminate scholarly ideas". Kerr (2006: 1127) illustrated how blogs provide "promising outlets for legal scholars interested in becoming public intellectuals". The importance assumed by law blogs in legal academia was further discussed by Duval (2018: 97), who argued that "blogs may often be a better vehicle to spread ideas in the digital era than traditional academic journals or edited volumes". In his article entitled "Publish (tweets and blogs) or perish? Legal academia in times of social media", Duval saw five main comparative advantages for blogs vis-à-vis traditional law reviews: access, speed, disintermediation, interactivity, and transparency.

Legal blogging has been considered as "public legal writing", as Murphy Romig (2015: 30) defines it. She emphasised the significant role that blogs play for lawyers who write them and "show a creative voice and distinctive personality, different from the client-driven stylistic practices of traditional legal writing" (2015: 66). This view supports Davis' (2020: 1175) claim that the digital public sphere is the way lawyers, who have a professional responsibility as citizen lawyers, are most likely to reach lay audiences and are encouraged to make public commentary, and think more deeply, about how arguments are constructed. The important role that the scholarly law blog plays in the practice of law, as a kind of bridge between legal academics and legal practitioners, finds an explanation in the impact that it has had on judicial opinions over the years. As Peoples (2010) shows in his study on the citation of blogs in judicial opinions, courts rely on the discussion of legal issues found on blogs to support judicial reasoning and analysis. This seems to suggest that law blogs embody credibility (Hyland 1999).

Recently, growing interest has been shown in the study of legal blogging from a linguistic perspective. Garzone (2014) explored the personal individualistic

https://doi.org/10.1515/9783111048789-005

dimension of law blogs or "blawgs", and likewise, Tessuto (2015: 85) investigated legal blogs as sites "for stance and engagement". Diani (2021, 2022) highlighted the argumentative dimension of law blogs and their dialogic function in post–comment threads. Attention has also been devoted to the importance assumed by law blogs as a tool for communicating legal issues with "the wider audience" of non-experts. Solly (2012: 52) examined a legal blog, *BabyBarista*, published first by *The Times* and then by *The Guardian* as "a fictional account of a junior barrister practising at the English Bar". Anesa (2018) discussed the "democratization of knowledge" by focusing on blawgs concerning environmental issues.

The present work contributes to this research strand by exploring the interpersonal language choices made by law bloggers who not only disseminate information to the audience, but who also express their stance towards the legal issues discussed. As defined by Hyland (2005a: 176), stance refers to "the ways writers present themselves and convey their judgements, opinions, and commitments". The authorial presence is marked by hedges, i.e., ways of withholding complete commitment to a proposition by means of epistemic modals (Hyland 2005b: 52), and complementary boosters, i.e., ways of strengthening a claim using adverbs (stance adjuncts) of the *surely, no doubt* kind. The other resources through which the writer marks his/her presence in the text are first person pronouns *I* and *we*, what Hyland (2001, 2005b) calls "self-mention", or, as recently rephrased in a study on digital genres, "the writers/speaker's intrusion in the text through use of first-person to emphasize their contribution" (Zou and Hyland 2022: 227). A further category of stance markers includes the whole set of attitude markers falling on the affective end of the epistemic-attitudinal continuum and marking the speaker/writer's more personal beliefs and feelings of surprise, frustration, agreement, importance, etc. (e.g., first person subject pronouns with verbs *hope, agree, prefer* as in *I hope/agree,* and adverbs such as *hopefully, surprisingly, unfortunately,* Hyland 2005b: 53).

In this chapter I analyse a corpus of law blog posts, written by law scholars commenting on legal cases relating to US and UK court decisions, to explore how stance features mark bloggers' presence and position them in relation to their arguments and audience. In the next section I provide a description of the corpus used for the study, as well as the methodology adopted. In sections 3 and 4, I report and discuss the results emerging from the analysis and, finally, in section 5 I draw some conclusions.

2 Materials and methods

The corpus used for this study consists of 96 posts (totalling 175,849 words), taken from two legal blog websites, selected over a six-year period, from 2017 to 2012 in the typical reverse chronological order of blog posts: one from the UK, United Kingdom Constitutional Law Association (UKCLA), and one from the US, SCOTUSblog (see Table 1). Only blog posts commenting on legal cases written by single-authored law professors and lecturers affiliated to UK and US universities were chosen. The law posts sampled from UKCLA are written by more than thirty different bloggers, six of whom write more than one post over the period. On the contrary, the posts selected from SCOTUSblog are written by the same four bloggers. The choice of single-authored posts was made in order to avoid potential differences with respect to the strength and focus of the positions expressed, and to create a more clearly identifiable category of blog. The two websites were selected for their relevance within the academic legal community. UKCLA "is the UK's national body of constitutional law scholars affiliated to the International Association of Constitutional Law". As stated in the 'About us' homepage, "its object is to encourage and promote the advancement of knowledge relating to United Kingdom constitutional law and the study of constitutions generally". Its blog, launched in 2010 (https://ukconstitutional law.org/blog/), is "a repository of expert comment and analysis on matters of constitutional law in the UK and further afield". SCOTUSblog (http://www.scotusblog.com/), founded in 2002 by attorneys Tom Golstein and Amy Howe, reports on cases of the Supreme Court of the United States. The blog posts are written by lawyers, law professors, and law students who analyse the cases before the US Supreme Court and post breaking news of court decisions.

Table 1: Composition of law blog corpus.

Blog	No. of posts	Tot. no. of tokens
UKCLA	46	98,201
SCOTUSblog	50	77,648
Total	96	175,849

From a methodological point of view, the corpus was searched for Hyland's (2005a, 2005b) stance features, as described in section 1 above, using AntConc (Anthony 2022) suite of programmes for electronic text analysis, in particular Word List and Concordance. A word list was generated from which stance markers with a frequency ≥ 10 were identified. All retrieved items were concordanced and manually checked to ensure that each performed the stance function it was assigned. To retrieve self-mention, the automated search was used, e.g., to separate instances of capitalised *I* vs. *i* by adjusting the Word List settings in AntConc (select "Treat case in sort"), then in the Concordance search (select both Words and Case and search for the capitalised form). However, the automated search returned some instances that needed to be discarded manually, e.g., *Section I.* In addition, some of the occurrences of *I* were eliminated because they are part of reported speech.

3 Stance features: A quantitative overview

A quantitative investigation of stance features in the corpus analysed provides the results summarised in Table 2. All frequency data reported in the table are presented as raw figures, followed by the normalised figure of the number of occurrences per 1,000 words.

Table 2: Overall frequencies of stance features in the corpus.

Stance features	Raw freq.	Freq. per 1,000 words (ptw)
Hedges	1,858	10.56
Boosters	398	2.26
Attitude markers	330	1.87
Self-mention	238	1.35

If we take an overview of the distribution of stance markers used in the corpus, undoubtedly the most striking feature is the heavy concentration of hedges (1,858 occurrences/10.56 ptw). This finding is not surprising as law bloggers give voice to their own positions through commentaries on court decisions, but this demands caution, as Zou and Hyland (2022) observe in their study of stance in academic blogs. The greater use of hedges reflects the need for law bloggers to avoid "the potential risks of over-assertion and attracting critical responses and disagreement [. . .] and so they err on the side of caution and pull their punches by hedging their arguments" (Zou and Hyland 2022: 232).

In order to persuade the readers of the validity of their argument, law bloggers make recourse to boosters that are the second most occurring stance option in the corpus (398 occurrences/2.26 ptw). However, as their small proportion reveals, they are cautious of using this feature as boosters are "a potentially hazardous strategy which runs the risk of attracting criticism and losing the support of an audience" (Zou and Hyland 2022: 233). The finding here echoes that of Zou and Hyland (2022:

233), who found that academic bloggers "are vulnerable to immediate and potentially caustic criticism and so closing down opportunities for disagreement with boosters is often avoided". This quantitative overview also indicates that bloggers express their attitudes to what they are discussing. Attitude markers, the third most frequent stance devices in the sample (330 occurrences/1.87 ptw), represent a contribution to the negotiation of knowledge and help bloggers gain community acceptance for their own views.

Previous research has shown that scholars use blogs "to construct their identity as competent academic bloggers and enhance their visibility in the blogging community" (Luzón 2012: 162). As Zou and Hyland postulate (2019: 724), "a strong authorial presence might be expected in academic blogs". However, our data reveal that the blogger's authorial stance is conveyed through other devices, such as boosters and attitude markers rather than self-mention, as Table 2 shows. Self-mention represents the least occurring stance feature (238 occurrences/1.35 ptw). This may be explained by the finalities of the blogs under investigation, which are managed by institutions. They focus on the construction of argumentation in the legal community and therefore present a more neutral stance of the blogger, rather than highlight what Bondi (2022: 8) calls the "biographical individual self", which is more evident in blogs that are managed by individual scholars. In the following section, each stance feature will be discussed in detail.

4 Stance markers: A qualitative analysis

4.1 Hedges and boosters

The most significant function of hedges displays the tendency for law bloggers "to downplay commitment to a proposition, allowing information to be presented as an opinion rather than accredited fact" (Hyland 2005a: 178). Table 3 shows the most frequent hedges used in the corpus with a frequency ≥10.

Table 3: Hedges in the corpus.

Hedges	Raw freq. ≥10	Freq. per 1,000 words (ptw)
would	651	3.70
may	287	1.63
could	276	1.56
might	206	1.17

Table 3 (continued)

Hedges	Raw freq. ≥10	Freq. per 1,000 words (ptw)
lemma seem*	174	0.98
likely	84	0.47
perhaps	75	0.42
rather	48	0.27
probably	21	0.11
somewhat	14	0.07
apparently	12	0.06
presumably	10	0.05

Epistemic modal verbs (*would, may, could, might*) are among the most frequently used hedging items in the corpus, totalling 1,420 instances of all hedges. By employing them, bloggers refrain from making straightforward assertions. For example, the modal verbs *would* and *may* in (1) can be taken to express the kind of tentativeness intended to hedge the evaluation (*unrealistic*) puts forth.

(1) Both these options *would* also require amendments to the CCP and *may* be unrealistic. (UKCLA)

Adverbs (*likely, perhaps, rather, probably, somewhat, apparently, presumably*) and the epistemic lexical verb *seem* complete the hedging resources found in the corpus. These adverbs signal that a certain degree of prudence is involved when the blogger states a certain opinion.

with so-called rights. This may	**seem** bizarre to those of us who believe
patibility. What might once have	**seemed** controversial has become run of
n 377. As it stands, both of these	**seem** difficult. In February 2014,
than other migrant workers. This	**seems** difficult to defend even from the
e against Abu Qatada in Jordan it	**seems** doubtful that even a higher
reaty without a statute, (and this	**seems** doubtful) that might not tell us
ny on the part of the judges which	**seems** inconsistent with establishing
in a number of cases in ways that	**seem** inconsistent. I myself have written
m is justified, it would certainly	**seem** odd to many Americans to say that

Figure 1: Concordances of *seem*.

The analysis shows a co-occurrence of negative words, mainly adjectives and verbs, with either the modal lexical verb *seem*, or downtoning adverbs, such as *perhaps, rather, somewhat* (e.g., *perhaps problematic, it rather ignores significant evidence, somewhat inconsistent arguments*), to reduce the force of the blogger's critical opinion. For reasons of space, only one expression may be discussed here by way of illustration: *seem** including the forms *seem, seems, seemed*. The concordance lines of *seem* are shown in Figure 1 below.

As we can see, the concordance is full of negative criticism that bloggers make less forceful through the modal lexical verb *seem*, used to weaken the threat of disagreement realised by adjectival criticism (e.g., *bizarre, controversial, difficult, doubtful, inconsistent, odd*).

While hedges tone down commitment or assertiveness, boosters indicate certainty (Hyland 2005a, 2005b). Thus, they are likely to be used when law bloggers feel confident about their opinions and express them openly. In fact, by using boosters they also add more emphasis to their statements. Probably, they expect their readers to agree with them on the topic discussed. Therefore, they use boosters to reinforce their beliefs and ideas. As Table 4 shows, adverbs are the most numerous boosting devices used in the corpus.

Table 4: Boosters in the corpus.

Boosters	Raw freq. ≥10	Freq. per 1,000 words (ptw)
indeed	105	0.59
clear/clearly	98	0.55
true	36	0.20
certainly	29	0.16
necessarily	29	0.16
of course	29	0.16
notably	16	0.09
no doubt	13	0.08
extremely	11	0.06
in fact	11	0.06
obviously	11	0.06
inevitably	10	0.05

Indeed, the first most frequent booster in the sample (105 occurrences), is often used to add emphasis to the statements made by bloggers or other legal actors represented in the posts. An example of such use can be observed in the following:

(2)   The fact that a particular doctrine is based on international comity does not mean that it must give a decisive role or *indeed* any role to the executive branch [. . .] (SCOTUSblog)

However, in (3) *indeed* is used to add a statement that supports the idea expressed in the previous sentence:

(3) Among the amici making this argument are the strange bedfellows The California League of Cities and the California Taxpayer's Association. *Indeed*, the City is represented by the Howard Jarvis Taxpayer's Foundation. (SCOTUSblog)

As shown in (4) below, the adverb *clearly* emphasises the force of the blogger's position and displays strong commitment to his speech act of disagreeing, which is overtly manifested through the stance adverb *strongly*.

(4) I *strongly* disagree, and believe that a facial challenge is improper because Section 5 is *clearly* permissible in federal elections. (SCOTUSblog)

Although *notably* occurs only sixteen times, it is a suitable example to discuss as it is used to argue a statement and persuade readers of a particular opinion. *Notably*, for instance, introduces an example considered representative of something that the blogger is discussing, as in (5). A different use of this adverb can be noted in (6), where it evaluates an action made by *Holyrood* as unusual and interesting.

(5) *Notably*, there is no requirement of standing to bring a complaint. The complainant need not have suffered discrimination. (UKCLA)

(6) Holyrood has *notably* already exercised this power in passing the Scottish Elections (Reduction of Voting Age) Act 2015, which allowed 16 and 17 year-olds to vote in the most recent set of Scottish elections. (UKCLA)

As we will see in the next section, boosters are often combined with attitude markers to enhance arguments.

4.2 Attitude markers

In analysing the data, we find that bloggers comment on court decisions using various strategies, ranging from statements averred by themselves, or attributed to a source introduced as distinct from themselves upon which they build a debate. They construct "a *persona* of disciplinary competence" (Hyland 2000: 125) through attitude markers.

As Table 5 shows, these markers mainly consist of items which indicate importance, significance, or surprise.

Table 5: Attitude markers in the corpus.

Attitude markers	Raw freq. ≥10	Freq. per 1,000 words (ptw)
important/importantly	115	0.65
significant/significantly	67	0.38
difficult	33	0.18
arguably	28	0.16
interesting/interestingly	27	0.15
surprising/surprisingly	17	0.09
unlawfully	12	0.07
unfortunately	11	0.06
helpful	10	0.05
remarkable/remarkably	10	0.05

Let us consider statements such as the following:

(7) Arlington v. FCC is *an extremely interesting decision* in that it has many of the trappings of an important Supreme Court ruling-a divided (5-1-3) Court [. . .] (SCOTUSblog)

(8) *Surprisingly little questioning* in the arguments centered on the Court's recent preemption decisions involving the US immigration laws. (SCOTUSblog)

From these extracts, it becomes immediately evident that it is the blogger that is heard in expressing subjective evaluation through the attitudinal adjective *interesting* or adverb *surprisingly*. This choice gives the blogger's statement objective status and, by implication, her/his claim is not intended to be challenged. The absence of any source for the claim suggests that the blogger accepts full responsibility for its validity and truth value. The blogger's strategy here is to present a categorical opinion that cannot be open to question. This leads us to suggest that bloggers are attempting to convince the academic community of their absolute opinion.

Interestingly, the adverb *unlawfully*, which explicitly expresses the writer's opinion toward an act that is considered unlawful, is mainly used by bloggers to report other people judging certain actions as unlawful. This can be seen in the following examples:

(9) Two appellants who had suffered a reduction in their HB under the 2012 regulations successfully argued that the bedroom criteria *unlawfully* discriminated against them contrary to Article 14 of the European Convention on Human Rights taken Article 1 of the First Protocol. (UKCLA)

(10) The EEOC then filed suit against Hosanna-Tabor, alleging that it had *unlaw-fully* fired Ms. Perich in retaliation for her assertion of her ADA rights. (SCOTUSblog)

Arguably often expresses a debatable opinion. In the corpus this adverb is used to refer more frequently to debatable facts or opinions that are expressed by others and not by the blogger himself/herself. Thus, by using this adverb, the blogger implicitly expresses his/her disagreement with the opinions and situations he/she reports, as the following examples taken from the corpus illustrate:

(11) The measure *arguably* imposed a tax on medical marijuana dispensaries and so the City argued that the measure must be put on the ballot at a general election, per the state constitutional rule governing the imposition of taxes. (SCOTUSblog)

(12) This conclusion is *arguably* correct, but equally the Order is silent on the matter and it is one about which again reasonable people could disagree. (UKCLA)

Another strategy which seems to gain relevance in the corpus, enabling the blogger to convey assessments without appearing overtly in the text, is the use of introductory *it* realised in two patterns: with a *to* infinitive and a *that*-clause (Francis, Hunston, and Manning 1998; Hunston and Sinclair 2000). Examples from the corpus are:

(13) Although the Court ended up resolving the two matters on relatively narrow grounds – disappointing some of the Justices as well as analysts – *it is important to* understand precisely what the Court did (and did not) hold in these two rulings, both of whose outcomes were decided by 8–1 votes. (SCOTUSblog)

(14) Given this emerging acceptance, *it seems hardly surprising that* the Supreme Court felt able to pronounce both that the Charter could be relied upon in UK law and that, in contrast with a breach of Convention rights, this meant that the provisions of the State Immunity Act 1978 could be disapplied, allowing both claimants to pursue the aspects of their claims which fell within the scope of EU law before an Employment Tribunal. (UKCLA)

On the surface the patterns are objective and impersonal. Here we can see the operation of what Halliday (1994: 32) calls "grammatical metaphor". Specifically, by putting the proposition *the Supreme Court felt able to pronounce* into a projection, the blogger can use the projecting clause, *it seems hardly surprising that,* to encode the objectivity explicitly. However, as Biber et al. (1999: 976) note, the function of this structure is "to express opinions and comment on and evaluate propositions in a way that allows the writer to remain in the background". This view also supports Hewings and Hewings' (2002: 370) claim that "the choice of it-clauses over a construction with a personal pronoun can [. . .] allow the writer to depersonalise opinions". The point I make here is that, although these patterns have an aura of objectivity, the choice of adjective, however, opens up a space which the blogger can use to construct his/her comment (*surprising, important*).

Another contributing factor to the subjectivity of these patterns is the presence of modifying adverbs before the adjective. Here subjectivity becomes particularly evident through the use of *hardly,* that seems to contribute to the expression of the blogger's strong personal commitment to the validity of his/her claim. In the examples above, we have seen that comments on legal cases are averred by the blogger without any signal of attribution to self. This implies more objectivity on the part of the blogger than if his, or her, self had been explicitly stated. In discussing this point, however, we have noted that, although there are claims whose pattern seems impersonal, there is no objectivity around them. This is the case with *it*-clauses where subjectivity becomes part of the claim through the choice of attitudinal adjectives, and even when the adjective is modified by adverbials. Thus, the use of impersonal statements gives only an initial impression of objectivity. Subjectivity is fairly overt.

4.3 Self-mention

Table 6 shows that self-mention accounts for the least proportion of total stance features in the corpus. As discussed elsewhere (Diani and Freddi forthcoming), this quantitative result reflects the communicative function of the law blogs under investigation. Law bloggers comment on legal cases with the purpose of

Table 6: Frequency distributions of *I, my, me* in the corpus (descending frequency order).

Self-mention	Raw freq.	Freq. per 1,000 words (ptw)
I	178	1.01
my	42	0.23
me	18	0.10

focusing on the ideational content of the court decision, the reasoning and legal argumentation, rather than self-expression closer to the communicative function of the early personal blog genre. This might explain why the first-person singular subject pronoun is not pervasive in the corpus, also confirmed by the very low frequencies of the object pronoun *me* and the possessive *my*.

To illustrate how *I* is used, the occurrences of this subject pronoun will be examined with the verbs that combine with it. Apart from the link-verbs *have* and *be* (*was, am*), the most substantial group of collocates accounts for verbs having to do with mental processes, such as *think, believe, doubt, suspect, found* (in the sense of *believe*), *understand.* The verb *think* is attested as the first most frequent collocate (13 occurrences), followed by *believe* (4 occ.), *doubt* (4 occ.), *suspect* (3 occ.), *found* (3 occ.). Examples from the corpus are:

(15)   *I think* the courts would be defeated, and in the end the Supreme Court would exercise its power under the Practice Statement of 1966 to reverse its position and reinstate the doctrine of parliamentary supremacy. (UKCLA)

(16)   In the space below, I analyze the merits portion of Mr. Clement's brief on behalf of the Arizona legislature, and point out why *I think* it fails to demonstrate that the IRC's creation and powers violate federal law. (UKCLA)

(17)   Justice Kennedy was surprisingly quiet at oral argument, asking just two questions about whether the identity of the party bound by the customary-international law norm should be part of the analysis at step one. The question is a good one, and *I believe* the answer is yes – the particular norm of customary international law must apply to a corporation before a corporation may be sued under the ATS for violating that norm. (SCOTUSblog)

According to Hyland (2000: 123), verbs such as *think, believe* allow writers to "build a personal ethos through an impression of certainty, assurance and conviction in the views expressed, an image strengthened with the use of personal pronouns". Examples (15) to (17) illustrate this certainty booster (to use Hyland's terminology) very well. Through *I think, I believe* the blogger shows an overt acceptance of personal responsibility for his/her argument and makes the conclusion stronger (*I think the courts would be defeated; I think it fails to demonstrate that . . .; The question is a good one, and I believe the answer is yes*). The second most frequent set of word-forms include verbs of saying or "discourse" verbs (Hyland 2005a) such as *discussed* (5 occurrences), *argue* (5 occ.), *offer* (5 occ.), *wrote* (5 occ.). As shown elsewhere (Diani 2021), the use of discourse verbs in the scholarly legal blog genre may be related to the discursive soft fields (including law in this case), as Hyland (2002: 126) noted for research articles:

The greater use of Discourse Act forms in the humanities and social sciences is more appropriate in an argument schema which more readily regards explicit interpretation, speculation, and complexity as accepted aspects of knowledge. [. . .] Writers in the soft disciplines therefore employed arguments that made greater use of Discourse Act forms which expedited the verbal exploration of such issues, facilitating qualitative arguments which rest on finely delineated interpretations and conceptualisations, rather than systematic scrutiny and precise measurement.

Through their posts, law bloggers try to gain credibility by establishing an appropriately "authorial persona" (Hyland 2001). This is done in the context of the argument that they construct. It is through this argument that they show authorial stance, as shown in the following examples taken from the corpus.

(18) Based on the paucity of commentary, it seems that few Supreme Court watchers or legal commentators followed the case. It appeared to present a rather narrow statutory interpretation question about the eligibility of a small group of children (children conceived through the use of assisted reproductive technology after the death of the sperm provider) to a particular federal benefit. *I argue*, however, that the Court's decision in this case may provide useful insights to how the Court may respond to some much broader and controversial arguments currently being asserted in the various same-sex marriage cases percolating through the court system. (SCOTUSblog)

(19) The President of the Supreme Court Lord Neuberger told 'The Times' last week that the idea is a 'recipe for complication, for cost and for unnecessary duplication', and the cross-bench peer Lord Pannick added that 'the proposal has no merit'. In this post, I am less kind to the idea. *I argue* that it is half-baked, unnecessary, and potentially dangerous, regardless of one's view of the desirability of an entrenched constitution protected by constitutional judicial review. (UKCLA)

As the examples show, the law blogger's role appears to be that of a writer who clearly assumes a role as arguer. Arguing about court decisions fuels the debate the blogger builds within his/her legal community. This result supports Kerr's (2006: 1127–1128) view that writing blog posts allows law scholars to participate in debates on law-related topics and become "public intellectuals", but also to create a niche for subject-matter experts. As claimed by Tessuto (2015: 103), they promote "critical debate and knowledge about the law and legal institutions" and advance "legal scholarship".

Although less frequent, another set of word-forms include modal verbs. The most frequently recurring ones being *will* (12 occurrences) and *would* (10 occurrences), respectively. *Will* is often used to direct the blogger's arguments and, as in (20), *will* is often followed by verbs such as *focus, explain,* and *consider*. Thus, it is often used to describe what the blogger will be discussing afterward in the post. The modal *would* is used frequently to hedge the blogger's ideas and arguments, as we can see in (21), where *would* is used to express a low degree of certainty.

(20) The Supreme Court will likely focus its ruling on the "ripeness" question as well, but – as *I will* explain below – questions of standing and ripeness are often tied up in complicated ways with the substantive question of whether a plaintiff has a winning constitutional claim on the merits. (SCOTUSblog)

(21) *I would* argue that the case does not concern parliamentary sovereignty but parliamentary legislative supremacy. (UKCLA)

Further evidence of the argumentative dimension of legal blogging comes from the analysis of the possessive *my*. It collocates with nouns such as *view* (6 occurrences), *mind* (3 occ.) which clearly indicate the adoption of the highly opinionated involvement of bloggers in discussing court decisions, as the following concordance lines show (Figure 2):

continue to be controversial.	**In my view**, it is not for the
ights demand. It is premised,	**in my view**, on the following two
The present case affords,	**to my mind**, a good example of the judic
a serious medical condition.	**To my mind**, Coleman is noteworthy not
es (who assert plausible, if	**to my mind** flawed, free speech

Figure 2: Concordances of *my* (right sorted).

As regards the object *me*, similar uses emerge in the corpus. When searching the collocates immediately following it, ranked by frequency (with a minimum frequency set at 2), the following three collocates appear on the list: *to me* (8 occurrences), *for me* (5 occ.), and *let* (1 occ.) as in *let me be clear*. No instances of *let me*-imperative are found. *To me* and *for me* occur in initial position to preface a comment. Here are some concordance lines from the corpus (Figure 3):

yet still the Court granted review.	**To me** that suggests a strong desire (by at
and the statute was too uncertain.	**To me,** the facts of these cases-and the
be the key to unlocking Walker?	**For me,** the critical fact in the case is

Figure 3: Concordances of *me* (left sorted).

Another case is the *it* introductory pattern + *seems* or *is* + adjective + *to me* + extraposed complement clause (Figure 4):

ased on the oral argument, it seems	**to me** that the Justices are closely
It seems like a very close call	**to me** but, if I had to guess, the
tarters, it is somewhat troubling	**to me** that a panel of the Commission
ot be sufficient but it was unclear	**to me** whether any of the justices agreed
just upside down."It was not clear	**to me** of his take on the standing questio.

Figure 4: Concordances of *me* (left sorted).

And also *me* as direct object in a clause (Figure 5):

The oral arguments have convinced	**me** that a majority of the Court will
governor." Indeed, what strikes	**me** most in reading the brief is that

Figure 5: Concordances of *me* (left sorted).

As the concordances show, law bloggers make themselves visible in their posts and present their own positions through commentaries expressing agreement and disagreement with court decisions, (i.e., *The present case affords, to my mind, a good example of . . .; it is somewhat troubling to me that a panel of the Commission found . . .*).

5 Conclusion

The study has shed light on how legal blog authors' rhetorical choices help create a particular stance and position them within the legal community. More specifically, the law blogs examined in this study reveal that bloggers prefer stance markers which downplay their commitment to the positions advanced in the posts. This result supports Zou and Hyland's (2022: 238) claim that academic bloggers "are conscious of opposing viewpoints and the potential for hostile criticism, which means their stance choices reflect a desire to present views in a less forceful way".

Interestingly, the analysis has shown that the blogger's authorial stance is conveyed through boosters and attitude markers rather than self-mention. This contrasts markedly with the original distinctive feature of blogs, i.e., individualistic self-expression. A possible explanation derives from the fact that the law blog websites are managed by an association (UKCLA) and professionals (SCOTUSblog) and have posts written by different authors who "may be influenced by the presence of the institution itself, taking responsibility for the web space that hosts the blog, and superimposing less personal forms of authoritativeness" (Bondi 2022: 17).

As Tessuto (2015: 103) observes, engaging in blogging "becomes part and parcel of a valuable, open access and makeshift system of scientific communication adapted to the needs and interests of disciplinary scholars". By analysing legal blogs, we hope to have contributed to a better understanding of the legal practices in web-mediated communication.

The study presented here is not without limitations, however. The corpus is limited in size and the analysis could be confirmed or disconfirmed by working on a different set of blogs. Further research may include the investigation of stance-taking by individual law blogs to be compared with the multi-authored institutional blogs examined in the present work. It would also be interesting to explore post–comment threads and to compare individual and multi-authored blogs.

Some implications can be drawn from this study for law teaching and legal scholarship. In an article entitled "Scholarship in action: The power, possibilities, and pitfalls for law professor blogs", Berman (2006: 21–23) sees "blogging as a valuable activity for law professors". This is because, as he points out,

Blog posts have provided the stimulus (and some text) for much of the "traditional" scholarship I have recently produced. Blogging has directly and indirectly played a role in a broad array of service opportunities and activities. Indeed, my most thoughtful posts or series of posts often at once serve as innovative teaching materials, an effective amicus brief, and the early first draft of part of a traditional article. [. . .] It is with such model in mind that I advocate blogging being seen – along with "serious legal scholarship" – as a preferred form of law professor activity.

As suggested elsewhere (Diani 2021), teaching modules on legal blogging for law students may contribute to developing the awareness and understanding of law blogs as a new digital genre of legal writing to be distinguished from traditional legal writing. In this light, the results of this study may help students become familiar with a range of linguistic strategies used to create stance and construct argumentation in blogging. Finally, along the lines of Luzón (2011), law blogs can be used, together with other digital genres such as online forums,

Twitter, etc., to help students map the argumentative practices and associate them with each specific genre.

References

Anesa, Patrizia. 2018. Popularization and democratization of knowledge through blawgs. *Iperstoria* 12. 155–168. https://doi.org/10.13136/2281-4582/2018.i12.400

Anesa, Patrizia & Louise Kulbicki. 2022. The impact of digitalization on legal communication: Introduction. *International Journal of Law, Language & Discourse* 10(2). 5–8.https://doi.org/10. 56498/1022022408

Anthony, Lawrence. 2022. *AntConc* (Version 4.1.4) [Computer Software]. Tokyo: Waseda University.

Berman, Douglas A. 2006. Scholarship in action: The power, possibilities, and pitfalls for law professor blogs. *Ohio State Public Law Working Paper* 65. 1–23. https://papers.ssrn.com/sol3/pa pers.cfm?abstract_id=898174

Biber, Douglas, Stig Johansson, Geoffrey Leech, Susan Conrad & Edward Finegan. 1999. *Longman Grammar of Spoken and Written English*. London: Longman.

Bondi, Marina. 2022. Dialogicity in individual and institutional scientific blogs. *Publications* 10(9). 1–20. https://doi.org/10.3390/publications10010009

Caron, Paul L. 2006. Are scholars better bloggers? Bloggership: How blogs are transforming legal scholarship. *Washington University Law Review* 84(5). 1025–1042.https://openscholarship.wustl. edu/law_lawreview/vol84/iss5/1

Davis, Kirsten K. 2020. [Classical] lawyers as [digital] public speakers: Classical rhetoric and lawyer digital public commentary. *Nevada Law Journal* 20(3). 1137–1176.

Diani, Giuliana. 2021. "In this post, I argue that . . .": Constructing argumentative discourse in scholarly law blog posts. *European Journal of English Studies* 25(3). 369–384. https://doi.org/10. 1080/13825577.2021.1988256

Diani, Giuliana. 2022. Managing discussions in law blogs: From post to comments. *International Journal of Law, Language & Discourse* 10(2). 9–21. https://doi.org/10.56498/1022022409

Diani, Giuliana & Maria Freddi. Forthcoming. Authorial stance and identity building in weblogs by law scholars and scientists. In Isabel Corona & Ramón Plo (eds.), *Digital Scientific Communication. Identity and Visibility in Research Dissemination*. London: Palgrave Macmillan.

Duval, Antoine. 2018. Publish (Tweets and blogs) or perish? Legal academia in times of social media. *Tilburg Law Review* 23(1). 91–108. https://doi.org/10.5334/tilr.4

Francis, Gill, Susan Hunston & Elizabeth Manning. 1998. *Collins COBUILD Grammar Patterns 2: Nouns and Adjectives*. London: HarperCollins Publishers.

Garzone, Giuliana. 2014. Investigating blawgs through corpus linguistics: Issues of generic integrity. In Maurizio Gotti & Davide S. Giannoni (eds.), *Corpus Analysis for Descriptive and Pedagogical Purposes: ESP Perspectives*, 167–188. Bern: Peter Lang.

Halliday, M.A.K. 1994. *An Introduction to Functional Grammar*, 2nd edn. London: Edward Arnold.

Hewings, Martin & Ann Hewings. 2002. "It is interesting to note that . . .": A comparative study of anticipatory 'It' in student and published writing. *English for Specific Purposes* 21(4). 367–383. https://doi.org/10.1016/S0889-4906(01)00016-3

Hunston, Susan &John Sinclair. 2000. A local grammar of evaluation. In Susan Hunston & Geoff
    Thompson (eds.), *Evaluation in Text: Authorial Stance and the Construction of Discourse*, 74–101.
    Oxford: Oxford University Press.
Hyland, Ken. 1999. Disciplinary discourses: Writer stance in research articles. In Christopher
    N. Candlin & Ken Hyland (eds.), *Writing: Texts, Processes and Practices*, 99–121. London:
    Continuum.
Hyland, Ken. 2000. *Disciplinary Discourses: Social Interactions in Academic Writing*. London: Longman.
Hyland, Ken. 2001. Humble servants of the discipline? Self-mention in research articles. *English for
    Specific Purposes* 20(3). 207–226. https://doi.org/10.1016/S0889-4906(00)00012-0
Hyland, Ken. 2002. Activity and evaluation: Reporting practices in academic writing. In John
    Flowerdew (ed.), *Academic Discourse*, 115–130. London: Longman.
Hyland, Ken. 2005a. Stance and engagement: A model of interaction in academic discourse. *Discourse
    Studies* 7(2). 173–192. http://dx.doi.org/10.1177/1461445605050365
Hyland, Ken. 2005b. *Metadiscourse: Exploring Interaction in Writing*. London: Continuum.
Kerr, Orin S. 2006. Blogs and legal academy. *Washington University Law Review* 84(5). 1127–1134.
Luzón, María José. 2011. "Interesting post, but I disagree": Social presence and antisocial behavior in
    academic weblogs. *Applied Linguistics* 32(5). 517–540. https://doi.org/10.1093/applin/amr021
Luzón, María José. 2012. "Your argument is wrong": A contribution to the study of evaluation in
    academic weblogs. *Text & Talk* 32(2). 145–165. https://doi.org/10.1515/text-2012-0008
Murphy Romig, Jennifer. 2015. Legal blogging and the rhetorical genre of public legal writing. *Legal
    Communication & Rhetoric: JALWD* 12. 29–81. https://ssrn.com/abstract=2679398
Peoples, Lee F. 2010. The citation of blogs in judicial opinions. *Tulane Journal of Technology &
    Intellectual Property* 13. 39–80.
Solly, Martin. 2012. Communicating with the wider audience: The case of a legal blog. *International
    Journal of Law, Language & Discourse* 2(1). 52–71.
Tessuto, Girolamo. 2015. "Posted by . . .": Scholarly legal blogs as part of academic discourse and
    site for stance and engagement. *Textus* 28(2). 85–107.
Zou, Hang & Ken Hyland. 2019. Reworking research: Interactions in academic articles and blogs.
    Discourse Studies 21(6). 713–733. https://doi.org/10.1177/1461445619866983.
Zou, Hang (Joanna) & Ken Hyland. 2022. Stance in academic blogs and three-minute theses.
    International Journal of Applied Linguistics 32(2). 225–240. https://doi.org/10.1111/ijal.12411.

Daniel Greineder & Dieter Stein
The Internet as a game changer in legal communication: Arbitration on the move

1 Purpose

In a programmatic statement Vogel (2019: 12) offered a definition of law in its multi-faceted nature that more or less enforces the perspective on law from changes in the medial[1] representation: "Law, then, is neither only an abstract logical system of norms nor only words on paper, but a specific style to communicate about and to negotiate the fundamental organization of society."

This being so, it is obligatory to look at how this "style of communication" is influenced by the nature of the medial media affordances in which this communication takes place. This topic will be approached from two complementary angles, that of a lawyer and that of a linguist, so there will be a double perspective on the issues. This contribution tries first to offer a general perspective on a mutual, if unidirectional, relationship between the law and its representation. The paper then provides an inside perspective on the way one particular legal genre is undergoing genre change under specific medial conditions.

2 Law

The larger issue of whether, and how, a major societal domain is affected by the new language media really points to two perspectives: to what "the law" actually is, and to the physical realization of "the law" in reality by actual acts of performing or applying the law through its real-life implementation.

There is "the law" as an abstract system of norms and cultural knowledge (which may or may not be textualized or scripted, as in Common and Code-based law), and there is the domain of law, not as a unitary entity, but as a societal field of activities, genres and discourse. Therefore, this genre perspective looks at the law in its realization through a number of genres in different language media, in the technical linguistic sense, such as wills, contracts, cross examinations, statutes, judgments and commentaries, and ways of bringing the law to citizens in a

---

1 For the purposes of this paper, "media" or "medial" are understood in a very circumscribed way to apply to spoken, written and Internet-based genres.

https://doi.org/10.1515/9783111048789-006

way that makes the law accessible to them. This distinction between the law as an abstract norm system, much like "langue" in linguistics, on the one hand, and, on the other hand, the manifestation or execution of the law in real time occurring in domain-specific actions, the genres, corresponds to two basic perspectives on the law: the law as an autonomous system or as a view of law viewed only as existing in effective actual acts of adjudication.

The notion of the societal domain is the coherence-defining level, underlain by the abstract system of norms and the cultural knowledge of those norms, that defines the coherence of legal genres under a top-down umbrella element and makes them functionally coherent.

However, the distinction between the abstract normative system and the realization in genres defines two very different approaches: the effect of what goes under the name of the "new medium" is surely different for the two following questions:

1 Is there an effect on the law as a normative, abstract system?
  and
2 Does the new medium have an effect on the system of genres?

For the purpose of the present discussion, this distinction leaves aside the logical possibility of there being an effect of a change in question number 2 on question number 1.

The first-mentioned effect, although broadly discussed for other domains of life, has hardly ever been discussed for law, even in such comprehensive comparative works such as Kischel (2019). It can be broken down into two sub-questions: is non-scripted law as a normative system different in kind and nature from scripted law? This issue must be left as a logical possibility, this not being the place to consider such a philosophical question.

The second aspect is a little closer to the issues covered in this volume. If legal norms are, in a historical developmental process, increasingly scripted, does this affect the system of legal norms? In other words, does a new way of "packaging" law in the form of "textualization", as seminally described by Peter Tiersma (2010), affect the normative substance? It is hard to speculate what Tiersma's answer would be, but it is still a fact that, even in a basic oral legal system, the adjudication of a new case still has to go back to dig out references to the normative system itself behind the precedent. Consequently, there seems to be no reason to assume that the mere writing down of the normative systems as written laws would make a difference to this system.

In addition, the process of textualization as described by Tiersma (2010) did, or does not, take place in a wholesale and exhaustive manner, affecting all and every form of law administration, but happened incrementally, in steps, and affecting different types of legal communication in different ways, in the process affecting the organization of legal practice in genres and the internal shape of the genres themselves.

3 Genre

Like any other domain, law is characterized by a complex landscape of functionally related genres and, as such, raises a number of issues of genre-theory. These genres are themselves related, in their typological pedigree, to genres from outside the field, e.g., the narrative or asymmetric spoken discourse, and the relationships with these outside categories is a matter of intense, theoretical interest. For instance, can a judicial narrative, or the narrative of a female in a harassment case really be considered narratives? Is a narrative given over the Internet still a narrative or is it simply an account? The definition of a genre hinges on its function in this domain. A standard question then is: are there new functions or are there only new ways of performing old functions?

The advent of the Internet has added a new chapter to a classic topic that has given us several volumes on medial-varietal linguistics (cf. Herring, Stein and Virtanen 2013 and Giltrow 2017): to what extent are written and spoken language different, and to what extent is electronic communication more written or spoken, or actually a new, third kind of language medium, with its own mechanisms of production, its own strategies of comprehension and cognitive processing and its own aura?

One of the more specific focuses has been the issue to what extent genres migrate to another language medium. Genre migration across medial boundaries is mostly conducive to a change in genre, such as a written diary to a blog. The blog is no longer a diary (Puschmann 2010). And, likewise, a chat exchange is no longer an oral conversation. It is also a truism that new language media generate new genres (Giltrow 2017, Giltrow and Stein 2009).

Yet, a specific challenge exists to linguistic genre theory insofar as in law there is a dogmatic kernel, where most legal activities are concerned with maintaining societal stability and ensuring recalcitrance towards change. This property is not shared by most other societal domains, such as science or sports.

4 Genre "statute"

We argued in §2 that the change in the representation of the normative system from abstract to written did not in itself effect a change in the norms themselves. However, it did affect the realization and practical application of the law in that the genre "statute", as a written genre, came into existence in its present form. If it is true that "the medium is the message", is there also a sense in which this could be true for the kernel layer of the legal domain, laws? D'Argenio and Sornicola (eds. forthcoming) deal specifically with the issue of how and why common, basically oral, law was supplemented by writing, codifying the law in written form. They point primarily to political factors with an effect on the medial realization of the substantive law or the doctrinal content of the law. The volume argues that, at the time, the intention was not to change the law, but to make a political statement by having the empowering authority of the written medium. The assumption of a special performative force of written statutes is another major step. As regards the law, the most important medially effected change described by Tiersma (2010) is "textualization". His central thesis is that the change from oral to written law does not simply imply a transposition into another physical medium (oral to written), but a change in interpretive processes: A legal document (a law, will, contract or judgment) is no longer "recorded" in writing, as a kind of an account (if at all), it is no longer "evidentiary" in character, but the written statute is now the law.

Textualization can be understood in a twofold way:
– An Act or Process effected from outside the text **making** it authoritative, and endowing it with performative authority
– From the side of the text: **assuming** authority

Textualization has, above all, led to a particular change in the interpretive process in the shape of the "literalness presumption": that is each word has been carefully chosen with the awareness that future judges will assume the words – in their plain meaning – to be understood as expressive of the lawmakers' intention.

Packaging the dogmatic substance of law in written form certainly makes law more explicit: it is an effect of the increasing "autonomy" of written discourse, with new interpretive conventions that are part and parcel of the genre. One might then be tempted to say that there is a significant modification in the genre "statute". However, this may not be the case as Common Law is precedent-based. It would appear that the very notion of a statute, or, more precisely, the genre statute, is medially tied to writtenness, and, in fact, to textualization. One cannot have a statute that does not involve the effects of textualization in the shape of the literalness presumption. Therefore, the codified, written statute with enactment force is surely such a new genre, and not a written account of the legal norms.

Written precursors of statutes, such as reports or descriptions of the law, were different genres, and not, as Tiersma stresses, statutes. They were evidentiary, not enacting. But there can be reports of statutes, or other kinds of genres derivative of statutes, such as interpretations, or explanations to the public. Referring to the age of the Internet, it is debatable whether there can be a fully-fledged statute in the digital medium in the genre sense, although there can be reports of statutes, or other kinds of genres derivative of statutes, such as interpretations, or explanations to the public.

The stasis, the fixity and long-term reliability of written and printed texts are factors that shape the specific, authoritative character of the written, textualized statute. And one is led to suspect that it arrests the further development of the law and its responses to the changing world. Although the "stare decisis" principle of common law is supposed to have the same effect, it is doubtful whether it can match the arresting and locking-in effect of the written form, and its supposed stability-inducing effect.

There are obvious advantages for other legal genres. Electronic format texts can be easily edited, amended and updated. Does that imply that, in the electronic format, there is no more "textualization" with authoritative documents, even if they are fully "integrated" in the technical sense? Is the "enactment" of legal provision affected? Does a will or a contract made out electronically receive legal force? Do we have an executable "will" if it is not handwritten, but computer-written and filed on a website?

Tiersma identifies several criterial aspects that are instrumental for textualization to occur:
- Actor-author (product of themselves)
- Writing down BEFORE the legal act/enactment occurs
- Author must INTEND the written text to be definitive (39)

All of these three aspects, defined for written language, may experience modification in an electronic "text" environment, in which even the very notion of the "text", as we prototypically associate it with written, static, fixed and invariant language, would or will have to be modified to be adapted to an electronic medial environment. What changes does the new medial environment make to the conditions for enactment?

Since genres are primarily defined pragmatically as activity types by their function in an institutional context, there can only be a "will" if the enactment force is preserved intact and unchanged. In other words: a contract would no longer be a contract if, for instance, the intention to enter a contractual relationship is not recognized. The "force" does not derive from linguistic, but pragmatic properties. This is also why "text-type" based attempts to define genres, by way of surface form characteristics, are doomed. Reconstruction of the "intention" is criterial in comprehension and processing. It may have all the same words and structures, but linguistic surface words do not make it the same genre, nor do medially induced structure changes themselves cause a change of genre.

5 Genre change in arbitration

The practitioner illustrating the changes mapped out by linguistic theory is faced with a twofold challenge. On the one hand, the subtleties of linguistic discussion of genre are so refined that any practical illustration runs the risk of sullying it; on the other hand, the legal practitioner's use of language is intense, continuous and largely unchecked by reflection, unless reflection is wrought by immediate practical expediency, such as the need to cross-examine a witness effectively. Such is the superficiality of some lawyers' use of language that solecisms like "the passive tense" are not unknown even in the High Court in London. The complexities of language use hold more surprises for the unsuspecting practitioner so that, like Moliere's character, he speaks in prose. The impact on legal genre in international arbitration may be unexamined within the profession but should offer rich pickings for linguists. A few examples are sketched here.

What is Arbitration? Arbitration is, at its simplest, a form of privatized litigation. Parties enter into a contract to opt out of the otherwise mandatory adjudication of courts and refer their dispute to resolution by a private arbitral tribunal, usually made up of lawyers, chosen to sit as a unique tribunal. Typically, arbitral tribunals resolve commercial disputes, mainly between companies or other commercial entities, but they also resolve disputes in sports law, public international law and, on occasion, employment law. It is not usual for arbitral tribunals to determine criminal or tax-related matters. The decision of the arbitral tribunal is normally a written award.

It is not uncommon for citizens to opt out of services provided by the state, and to make arrangements by private means. In many countries, citizens may choose private over public health care, private over state schools, and private over public pension plans. In arbitration, as in other cases, they do so within a state framework. The two most important elements are (1) national arbitration legislation, and (2) the New York Convention that provides an almost global framework for the recognition of arbitral awards internationally.

Arbitration proves attractive to the linguist for several reasons. It is a common and arguably the most popular choice for resolving disputes between international parties. For example, a Turkish and a German construction contractor may enter into an agreement to build a bridge in Sub-Saharan Africa. It is unlikely either contractor would choose to refer disputes to local courts in Africa, and equally each would suspect the other of benefiting from an unfair advantage in its own jurisdiction.

Arbitration offers a relatively neutral forum and thus brings together parties, counsel and arbitrators from all over the world. It is exposed to many languages and trends in global communication and miscommunication. Further, it is relatively well-budgeted. Parties in larger cases often have the resources to work with international counsel, to travel and to try out new technology. Finally, it is anomalous in the legal world insofar as it does not have a language of its own. Parties choose the common language of the proceedings, according to their preferences, even though that language may be different from the language of the legal texts applicable to the case. For example, even where the main contract in the proceedings is governed by German law, the parties may choose to conduct the proceedings in English because that is the language of their business cooperation.

Although, as most other lawyers, arbitration lawyers can be somewhat technophobic, they have adopted some of the new media. This article considers three trends.

6 The textualization of the hearing

Arbitration advocacy is seldom theatrical, if "theatrical" connotes speeches grandly declaimed in a histrionic style. The hearing itself is just a procedural step, not a great day of reckoning and is typically preceded by a series of written submissions in which the parties have set out their views. Moreover, the sedate atmosphere of a hotel conference suite or purpose-built hearing centre tends to stifle grandiloquence. And yet, restrained though the cross-examination of witnesses is, often involving an interpreter, the move to virtual hearings, especially post-COVID 19, and heavy reliance on court reporting services, suggest that the textualization of the hearing increases.

Where parties, counsel and arbitrators are often based in different parts of the world, travelling to hearings in typical arbitration venues, such as London, Geneva or Singapore, is often expensive and inconvenient. However, it was the COVID 19-related lockdown that provided the crucial fillip to embrace virtual hearings. These may use regular meeting software, such as Zoom, Teams and

Webex, as well as other specialist packages, such as those used by Opus 2. They allow for virtual breakout rooms for clients to hold confidential meetings with their lawyers, the sharing of documents, and the recording and almost instant transcription of the record, which participants can follow as if reading subtitles.

The immediacy of the proceedings in virtual hearings is deadened. Television and cinema may use techniques to overcome the remoteness of a screen performance, compared with the experience of live theatre, but general advice to arbitrators on facial expressions and lighting does little to restore that urgency. Speakers appear on screen, discussions take place on split screens, and the text rolls on like a news ticker. Live transcription services are not new, nor has it been unusual in recent years for one or other witness to testify online. What is new is the departure from the real room and the immediacy of real people. Graphs, diagrams and documents gain prominence with the effect that a hearing becomes textual and the live human voice recedes. In dramatic terms, the effect is one step away from performance *viva voce* to the *Lesedrama,* the play that is better read quietly at home rather than experienced on stage.

The hearing becomes a *Lesedrama* when, some months later, the arbitrators review the transcript and re-evaluate the testimony with only a residual impression of the diction and manner of the witnesses. They may do so with the assistance of Post-Hearing Briefs in which the parties draw conclusions from the hearing and instrumentalize its progress to their own ends. The question then is not so much about what happened at the hearing, but rather as to who can impose the dominant interpretation.

An important first effect is the decentering of the Arbitration Memorial. The changes to the memorial or brief brought on by the new media stand in chiasmic relation to the textualization of the arbitration hearing. The memorial or brief, sometimes just a "written submission", owes something to the civil law court tradition of the *écriture* or *Schriftsatz* in which litigants or, more usually, their lawyers set out their cases in some detail for the judge to read. Although arbitration, with its cross-examination and oral hearings, owes much to the common law tradition, modern international practice draws on civil law culture as well. Aside from the mixed legal sources of arbitration, it is useful for arbitrators, who may, in a three-man tribunal, be spread across the world with one in Geneva, another in Singapore and a third in Vancouver, to have something to read in their own time, and without the inconvenience of having to schedule hearings.

The typical memorials, Statement of Claim, Statement of Defence, Reply and Rejoinder, may run to tens, sometimes hundreds, of pages and they ought to contain the full statement of a party's position. The lawyers drafting them will draw together facts, such as documentary evidence, oral evidence, usually contained in witness statements, and the findings of expert reports commissioned for the purpose of the arbitration. Although memorials are often multi-authored, law firms prefer to impose a house style ensuring a high degree of consistency.

Given the occasionally chaotic nature of law firm life, drafting memorials has long depended on word processing software, usually Microsoft Word. A number of developments have brought further change. Memorials can be very long and depend on extensive footnoting and cross referencing. Law firms increasingly provide them to arbitrators and their opposing counsel with a hyperlinked version of the memorials. It is no longer necessary to check the document referred to in a footnote by looking for it in a hard copy file. Rather, a reader need only click on the link to be taken straight to the document.

This has obvious advantages for checking references. It also allows for documents to be transferred across the globe more easily. If arbitrators work from electronic files, they no longer need hard copies, which can be expensive and cumbersome to transport. Equally, if perhaps more questionably, electronic files are said to reduce the environmental harm of arbitration cases. It is no longer necessary to print, say, 20,000 pages, only to discard them a month or two after the conclusion of the proceedings.

The medial implications of this development have so far gone unnoticed. Readers will experience a memorial that is linked to sometimes hundreds of other documents in a very different way to a text whose structure and line of argument are determined by the author. One can, of course, sit down and read a hyperlinked memorial from start to finish showing deference to the authorial voice. Yet, it invites exploration and, as one foray into the expert reports leads readers to observe the graphs and tables in the annex to that report, the authorial authority is eroded. In making their reasoning transparent, and by opening up every document to the immediate review of the reader, the authors have ceded some of their authority over the text.

7 The public award as a new genre

To understand this third aspect of the impact of the new media on arbitration it is necessary to introduce arbitral institutions, such as the International Chamber of Commerce International Court of Arbitration, the London Court of International Arbitration or the Singapore International Arbitration Centre. Arbitral institutions are not courts. Their main services are easily summarized, but the extent of their control over the process is harder to define. They provide administrative services in support of arbitration which include managing the arbitrators' fees, providing some extra rules of procedure and sometimes scrutinizing the award. It is a matter of debate whether arbitral institutions have too much power. Are they impresarios who enable the parties to put on a show, i.e. the arbitration, or are they more like directors who determine the course of the proceedings, perhaps even by imposing some form of Regietheater on the parties? We need not answer the question here. Suffice it to say that institutions wield power and are in competition with each other.

In recent years, a number of institutions have encouraged the publication of arbitral awards. This has already been happening for a long time in relation to awards in investor-state disputes rendered under the auspices of the International Centre for Settlement of Investment. Reasons for giving greater publicity to the greater dissemination of awards include giving more publicity to the institution, increasing the accountability and transparency of the process – public as opposed to private justice – and inclusivity. On the last point, arbitration is sometimes seen as a "closed shop" where insiders keep inside knowledge for themselves. Giving outsiders wider access to information would supposedly open up that closed shop to more diverse users.

Neither these goals nor the desire to achieve them depends on new media. However, the new media, notably the internet, facilitate this initiative and it would be hard to imagine its success without easy internet access. It is doubtful whether the legal and linguistic ramifications are understood. In substance, a private text will become a public one. An award, of which perhaps six or eight copies might previously have existed, becomes a public document.

Fundamentally, an arbitration is a private contractual process intended to determine the disputes between two parties. It is subject to minimal standards of judicial oversight, but it is not a precedent and does not establish a precedent in the sense of a court judgment in the common law legal system. Arbitrators write principally for the parties. If the decision is one that the parties are content with, and the award meets minimal legal standards, then it does not matter if the award is of wider use as the discussion of a legal problem.

It remains to be seen whether arbitrators will respond to this greater publicity. Will stylistic traits become a form of branding? Will they adapt their style to a wider audience? And will they change their reasoning to establish influential, if not precedentially binding, forms of reasoning?

8 Conclusion

Obviously, the processes described here amount to significant changes in the realizational mode of the law. Furthermore, it is tempting to undertake a comparison with language, as has been shown briefly above. In modern linguistic change theory, there is the idea that there are what might be called "subsystemic" changes, such that groups or networks change local aspects of the language under certain communicative conditions, reinforced by the fast modes of communication through Internet genres, such as blogs, twitter or Instagram. They do not only represent new genres, but also drive innovative uses of linguistic forms. The totality of such changes might eventually result in what – on a higher level of the totality of the language – might register as "a change" of language. Therefore, it would appear that the effect of the Internet is the same in the bottom-up realization of both systems of norms: subsystemic innovation and acceleration in the realization process, whether or not changes eventually result in the higher-abstraction systems.

References

Bhatia, K. Vijay & Mauizio Gotti, Azirah Hashim, Philip Koh, Sundra Rajoo (eds.). 2017. *International Arbitration Discourse and Practices in Asia*. London: Routledge.

Breeze, Ruth. 2019. The practice of the Law Across Modes and Media. Exploring the Challenges and Opportunities for Legal Linguistics. In Friedemann Vogel (ed.). *Legal Linguistics Beyond Borders: Language and Law in a World of Media, Globalisation and Social Conflicts, Relaunching the International Language and Law Association (ILLA)*, 291–314. Berlin: Duncker & Humblot.

D'Argenio, Elisa & Rosanna Sornicola. Forthcoming. *Alla conquista del codice scritto. L'apporto dei visigoti e dei longobardi alla formazione della lingua del diritto europeo*. Berlin: Mouton De Gruyter, Foundations in Language and Law Series.

Giltrow, Janet & Dieter Stein. 2009. *Genres in the Internet*. Amsterdam: Benjamins.

Giltrow, Janet. 2017. Bridge to genre: spanning technological change. In Carolyn Miller, & Ashley Kelly (eds.). *Emerging genres in new media environments*, 29–62. London: Palgrave.

Herring, Susan, Dieter Stein & Tuija Virtanen (eds.). 2013. *Pragmatics of Computer-Mediated Communication*, Berlin: De Gruyter Mouton.

Kischel, Uwe. 2019. *Comparative Law*. Oxford: Oxford University Press.

Puschmann, Cornelius. 2010. *The corporate blog as an emerging genre of computer-mediated communication: features, constraints, discourse situations*. Göttingen: Universitätsverlag Göttingen.

Tiersma, Peter. 2010. *Parchment, paper, pixels: Law and the technologies of communication*. Chicago: The University of Chicago Press.

Vogel, Friedemann (ed.). 2019. *Legal Linguistics Beyond Borders: Language and Law in a World of Media, Globalisation and Social Conflicts, Relaunching the International Language and Law Association (ILLA)*. Berlin: Duncker & Humblot.

Jekaterina Nikitina
Linguistic markers of participation in international criminal trials

1 Introduction

Modern criminal proceedings, especially at the international level, devote significant attention to the criteria of procedural fairness, including respect for the dignity of the accused and effective participation (Rossner and Tait 2021). But what counts as effective participation? International legal sources frame it as "access to justice" (Art. 8, Universal Declaration of Human Rights), "right to a fair trial" (Art. 6 (1), European Convention on Human Rights), and right to a "fair and expeditious trial" (Art. 64 (2), Rome Statute), putting emphasis on the general setting. If the notion of effective participation is considered from a strictly linguistic viewpoint, it can be defined through Goffman's (1974; 1981) concept of a "participation framework", according to which recognized, or "ratified", participants are able to interact within a communicative event, thus putting the effectiveness of participation, with the rights and obligations to speak and be heard, on the same plane. Video-link, remote or virtual participation in courtroom interaction may exacerbate the challenges inherent in ensuring effective participation (Rossner and Tait 2021). Although the remote dimension of courtroom interaction, with the respective ethical and legal challenges, is not a new phenomenon (Mikkelson 2016), it has been rekindled with the COVID-19 pandemic (International Commission of Jurists 2020: 6; Rossner and Tait 2021), inviting research into the "participation framework" (Goffman 1974; 1981) in increasingly digitalized courtroom interaction.

This study offers a sociolinguistic corpus-assisted discourse analysis of the trial transcripts in the Situation in Central African Republic II (CAR II), *The Prosecutor v. Alfred Yekatom and Patrice-Edouard Ngaïssona ICC-01/14-01/18* at the International Criminal Court, aimed to assess the linguistic realization of participation. Who are ratified participants in the proceedings? What are the terms of reference used to refer to different participants, especially to those belonging to vulnerable categories, and do these terms change when used by different actors? Does a different participation mode – in-person or remote – affect the linguistic realization of participation?

To set the theoretical framework, Goffman's "participation framework" (1974) is outlined and supplemented by linguistic insights into deixis/indexicality as markers of participation (Wortham 1996; Felton Rosulek 2015). Courtroom interaction is here conceptualized using Goffman's interactionist notions (1981), which involve hierarchical context of communication and strict turn-taking rules (Section 2). As every

https://doi.org/10.1515/9783111048789-007

international court, the ICC operates in a complex legal, social and linguistic context, briefly described in Section 3, along with the study materials. Section 4 offers the sociolinguistic analysis of the participation framework against the background of procedural digitalization, followed by discussion and conclusions in Section 5.

2 Participation and its linguistic markers

This study draws on Goffman's concept of "participation framework" (1974; 1981), i.e. the recognition of the participant's status based on their ability or inability to act within a communicative event, as well as their rights and obligations to do so in a certain way. Along these general lines, participants to a given activity may be either ratified, i.e. with their status recognized, or unratified, i.e. with their participation status unrecognized before the event. Once it has been established whether participants are considered as ratified or not, it becomes possible to describe their function or role (Goffman 1981: 137), i.e. listener or speaker, and all their contextual variants.

In a courtroom setting, it is virtually impossible to imagine an unratified speaker, on account of the highly institutionalized and hierarchical context of communication and strict turn-taking rules. On the other hand, listeners may be subdivided into three categories (Goffman 1981: 9–10):

– unratified listeners who "overhear" conversation inadvertently or on purpose, e.g. anyone in the public gallery;
– ratified unaddressed listeners, who are ratified participants, but are not specifically addressed at a given moment of proceedings, e.g. a public prosecutor or a judge during a witness examination by the defence counsel; and
– ratified addressed listeners, who are expected to take the next turn, e.g. a witness after a lawyer or a judge ask them a question.

Contextualization or "context-sensitivity" (Sacks et al. 1974: 699) is key to understanding the participation framework (Goffman 1981: 140), and, by extension, the communicative goal that is to be achieved. Courtroom interaction is "institutional talk" (Drew and Heritage 1992) and "more-than-two-person talk" (Sacks et al. 1974: 696–735). Courtroom talk-in-interaction is highly sequential and is organized dialogically in multiple adjacency pairs, or questions from the institutional party (e.g. a judge or a lawyer) and answers from a lay party (e.g. a witness or a defendant), resulting thus in "ritual interchange" (Goffman 1981: 17, cf. Goffman 1967: 19–22). In other words, on account of the institutional setting, courtroom interaction occurs as naturally expected sets of turns, with a number of "ritual constraints" (Goffman 1981: 19–20) on what is allowed and what is not. For example, in a courtroom a discursive ritual constraint would be the use of titles, roles and surnames rather than first names (O'Barr 1982), whereas a purely physical one would be the requirement to stand up when addressing the court. This regularly happens within established communities of practice (Eckert and McConnell-Ginet 1992), where certain social norms and conventions are shared, and expected to be implemented, even when there are no written requirements to do so.

This study concentrates on the linguistic markers of the participation status realized through deixis/indexicality (Biber et al. 1999; Morford 1997). Whenever the participant's name and role are verbally acknowledged, either through vocatives (cf. Wortham 1996) or overt references, their participation status is ratified, or rather their identity as a ratified participant is indexed. The study overviews multiple terms of reference for the participation framework, as well as their deictical representation. Deixis in courtroom interaction has been conceptualized as a powerful framing tool which can, for example, enable a lawyer to make the jurors "see" something in a given light through deictical indicators. Deixis serves also as a marker of a participation status (Felton Rosulek 2015: 156), for example, by using pronouns, which "do not have a set meaning prior to their use" (Felton Rosulek 2015: 34; cf. Biber et al. 1999; Morford 1997) and can "both refer to and establish an interactional group" (Wortham 1996: 332). In other words, they serve the purposes of ratifying a participation status which was not necessarily ratified before the use of the pronoun (Felton Rosulek 2015: 34; cf. Morford 1997). For example, in an international courtroom, ratified speakers, such as the presiding judge or the party with the active turn, may use vocatives and second-person pronouns to define who the addressees are, thus making them ratified addressed listeners who are expected to become ratified speakers at the next turn. By contrast, when the speaker uses a first-person pronoun, he or she "anchors himself or herself (and sometimes others) as a character in the discourse and can mark his or her level of responsibility for the message being conveyed" (Felton Rosulek 2015: 156).

Besides being directly addressed through vocatives and second-person pronouns, participants can also be referred to in discourse in the third person. To do so, they need to be linguistically represented in text and talk. For what constitutes lexical representation, the study relies on van Leeuwen's (2002) framework for "the representation of social actors", according to which it is possible to distinguish between personal and impersonal representation. The first category includes *nomination*, i.e. the use of participant's "unique identity" (van Leeuwen 2002: 322), which is quintessentially represented by their name. Another personal representation type is *categorization*, which can be subdivided into *functionalization* (reference to the participant's role in the context, e.g. soldier, defendant) and *identification*, i.e. reference to "what [participants], more or less permanently, or unavoidably are", e.g. colleague (van Leeuwen 2002: 324). *Impersonalization* (van Leeuwen 2002: 239), on the other hand, omits or replaces the participant's unique identity with different qualities, including *spatialization* (e.g. referring to the country of origin rather than the person), *utterance autonomization*, (i.e. participant's representation by reference to his/her utterances), *somatization* (i.e. referring to the participants by mentioning his/her body parts, e.g. "his hands", etc) and other phenomena that background the identity of participants.

3 Study design

This section overviews the communicative context of the International Criminal Court (ICC) in 3.1, moving on to the description of materials in 3.2.

3.1 The International Criminal Court

The International Criminal Court (ICC) is a judiciary body of the United Nations based in The Hague and established in 2002. Its jurisdiction is defined by the Rome Statute to cover the most serious crimes, i.e. genocide, crimes against humanity, war crimes, and (as of 17/07/2018) crime of aggression. In contrast to many other ad hoc international criminal tribunals, such as the International Criminal Tribunal for the former Yugoslavia (ICTY), or the International Criminal Tribunal for Rwanda (ICTR), the ICC is a permanent, autonomous court (ICC 2020: 10).

As every international court, the ICC operates in a multilingual context, both in terms of its jurisdiction and its personnel (Swigart 2019: 2). The official languages of the Court are those of the United Nations: Arabic, Russian, Spanish, English, French and Chinese (Art. 50, Rome Statute), whereas its working languages, i.e. those employed for daily administration of its activities, are English and French. In addition to which, the Court has to accommodate more than thirty "situation languages",[1] i.e. languages predominantly coming from the African continent that are used in ICC investigations, trials, outreach activities, etc., and which fall under the category of "languages of lesser diffusion" (Salaets et al. 2016). Daily operation of the ICC largely relies on linguistic services that, alas, remain frequently behind the scenes, "unsung and unseen" (Swigart 2019), and while ICC

---

1 The term "situation" is used by the ICC to refer to "temporal, territorial and in some cases personal parameters", whereas the term "case" is applied to "specific incidents within a given 'situation' during which one or more crimes within the jurisdiction of the Court may have been committed" (ICC 2016: 3, see also Swigart 2019).

interpreters "have given voice to the voiceless" (ICC interpreter, quoted in Swigart 2019: 4), it remains an open question whether their own voice is counted as a ratified participant in courtroom proceedings or not.

Before continuing with the analysis, a clarification is needed of what type of courtroom proceedings characterize the ICC. There is no dearth of research on how different legal traditions – civil law and common law – may impact the material and linguistic course of the proceedings. This distinction takes on importance in terms of the roles and functions of the participants, in that common law (adversarial) proceedings grant vaster procedural independence to the parties on evidentiary truth-finding matters, limiting the judge's role to that of a (predominantly) ratified listener and an "arbiter of procedural matters" rather than an active participant in the truth-finding process (Schmitt 2021: 486). Civil law (inquisitorial) systems, on the contrary, do not hand over the reins of the truth-finding process to the parties. In a civil law tradition, the presiding judge acts as a "director" of the process: he or she "determines which evidence to examine, calls the witnesses, and questions them as well as the accused" (Schmitt 2021: 486), whereas the role of the counsel is significantly relegated. The ICC procedure is represented by a complex "blending" or "clash" of civil and common law traditions (Schmitt 2021: 486). Judge Bertram Schmitt, quoted above in his "broad-brush synopsis" (Schmitt 2021: 487) of legal traditions, is a legal scholar from Germany and an active judge of the ICC who possesses an analytically fine-grained understanding of different procedural participation issues in the ICC context (Schmitt 2021). The practical and academic awareness that Judge Schmitt has about his role in a blended international jurisdiction of the ICC was an additional criterion in the choice of materials for this study, as Judge Schmitt acts as a presiding judge in case ICC-01/14-01/18.

3.2 Study materials

Another criterion for the choice of case study was temporal, in that I wanted to analyse a trial which took place during the active phase of the COVID-19 pandemic, where some parts of the proceedings were reasonably expected to be digitalized and conducted remotely. The terms "remote participation" or "video-enabled hearing" (Rossner and Tait 2021: 3), frequently employed in the literature, include different degrees of digitalized participation, which can be

a)   *skewed*, when only the defendant is video-linked from a remote location;

b)   or follow the "distributed court" pattern (Rossner and Tait 2021: 2), when participants are video-linked from multiple locations, typically their homes and offices, and the judicial staff is connected from an otherwise vacant courtroom, which happened during the pandemic in most domestic settings, or

c) *blended*, to borrow a term from the conference settings, when a reasonably balanced amount of participants attend an event from their home or office and a similarly reasonable amount of participants attend it from the courtroom.

Case ICC-01/14-01/18, *The Prosecutor v. Alfred Yekatom and Patrice-Edouard Ngaïssona*, was chosen as an example of a blended courtroom interaction.

The case concerns crimes against humanity and war crimes allegedly committed between December 2013 and December 2014 in the Central African Republic (CAR). The case deals with an armed conflict between two organised armed groups, the Séléka and the Anti-Balaka, within which the latter carried out "a widespread attack against the Muslim civilian population, perceived – on the basis of their religious or ethnic affiliation – as complicit with, or supportive of the Séléka and therefore collectively responsible for the crimes allegedly committed by them" (ICC case info 2021). The defendants on trial held positions of power during the Anti-Balaka regime and thus were being tried for their contributions to the abovementioned crimes. Both had been in ICC custody since December 2018-January 2019 so, at the time of active proceedings, both were physically present in The Hague. The trial opened on February 16, 2021, i.e. during the second wave of pandemic-related restrictions, when travel between the CAR and the Netherlands was hindered, making it difficult for some participants to attend the trial in person.

At the time of the corpus collection and analysis (November 2021), the Prosecution's presentation of evidence was still ongoing.

Table 1 summarizes the texts included in the corpus. All texts were downloaded from the official ICC website https://www.icc-cpi.int/, by filtering for the relevant case and then going to the Transcripts section. Many parts of evidentiary hearings were redacted; some transcripts were therefore either not available or were heavily sanitized.

LancsBox 3.0 software was used for text search and quantitative overview in the spirit of corpus-assisted discourse analysis (Baker et al. 2008; Partington et al. 2004), building on the synergy between quantitative data analysis and (Critical) Discourse Analysis (van Leeuwen 2002), as applied to issues of linguistic representation. The quantitative data are presented as absolute frequencies (AF) for the whole corpus and as relative frequencies normalized to 10,000 words (NF) using MS Excel sheets for the analysis of opening statements by different participants, to assess how different actors use the same terms of reference. Even though there were two defendants on trial, Yekatom's defence team did not make their opening statements, postponing them to a later period of trial, so these data are absent during the current analysis. A simple range is indicated, too, showing in how many texts an item occurs in the corpus, expressed as a percentage.

Table 1: Corpus composition.

Phase	Trial: Opening statements	Trial: evidence (March)	Trial: evidence (April-May)	Trial: evidence (June-July-August)	Trial: evidence (Sept-Oct-Nov)	Total
Days	16/02; 17/02; 18/02/2021	15/03; 16/03; 19/03; 24-25-26/03; 30/03/2021	26/04; 28-29-30/04; 10-11-12/05; 24-26/05/21	02-03-04/06; 07-08/06; 05-09/07; 12-15/07; 21/07; 30-31/08/21	01/09; 27-28-29/09; 02-03-04/11/21	45
No of transcripts	3	7	10	18	7	45
Tokens (transcript)	70,743	125,687	207,927	299,152	86,771	790,274

4 Findings

Because of the pandemic-imposed limitations, many teams attended hearings on a rotation, and the usual ICC courtroom ritual, requiring all parties to nominate the active participants at a given trial stage for the record, acquired a new importance, even more so, considering that some participants attended the court via video-link. On the first trial day the remote participants were not given the floor by their in-person present colleagues, and the camera did not provide a shot of their faces, thus effectively deratifying them of both their voice and their face. This was corrected on the following day (1).

(1)   MS MASSIDDA: (Interpretation) Good morning, President. Good morning, your Honours. *I'll see if I can do better than yesterday in terms of introducing the members of the team.* [. . .] In Dakar, I think that *my colleague would like to introduce himself.*
      MR FALL: (Via video link)(Interpretation) Good morning, your Honours. *I am indeed one of the members* of the Legal Representatives for Victims and *I will be attending these proceedings from this office in Dakar.* Your Honours, please allow me to speak while seated. I'm quite a tall gentleman and if I do stand up, I think it will be difficult for the camera to catch my image correctly. So if you would be so good as to allow me to address the Court seated. Thank you.

> PRESIDING JUDGE SCHMITT: Of course we do. We prefer that we see you, and this is above formality, of course. [17/02/21]

As example (1) illustrates, the blended participation mode elicited the need to change some of the courtroom rituals, i.e. allowing the remotely-linked participant to remain seated while addressing the court, confirming the importance of face-to-face interaction and the value of the virtual face as a marker of status ratification. This novel ritual was confirmed on multiple occasions when other counsel and witnesses joined via video-link.

The way different participants are lexically referred to shows how different aspects of their identity are indexed, including "the social groups to which the speaker thinks they belong" (Wortham and Locher 1996: 562, cited in Felton Rosulek 2015: 56). In general, a strong trend towards personalization was observed. Every participant's identity, with the exception of protected witnesses, was anchored to the reality of proceedings through the initial ritual *nomination*, typically including an honorific or title (Mr, Mrs, Maître, etc) and the surname ("Mr. Yekatom"), followed or preceded by *functionalization* ("lead counsel", "legal assistant", etc).

The next paragraphs address the lexical representation of different groups of participants: the defendants (4.1); institutional participants (4.2); witnesses and victims (4.3); and interpreters (4.4).

4.1 Defendants

Table 2 reports terms of reference used for the defendants. Their full formal names ("Mr Patrice-Edouard Ngaïssona", "Mr Alfred Yekatom") were used to announce the case by the Court officials. Throughout the proceedings, when they were addressed or otherwise indicated, the prevalent pattern of reference was an honorific "Mr" followed by the surname, which represents the most formal type of nomination (Felton Rosulek 2015: 57). Peculiarly, this highly standardized formula was changed at a later stage of proceedings, adding to Alfred Yekatom's name his nickname, *Rombhot*, at times spelled as *Rambo*, used by the witnesses (2).

(2) I think this was not a secret. It is something that was known to all Central Africans. *Rambo*, he was well-known. And when I talk about *Rambo*, it is *Yekatom* I'm referring to. [Witness, 16/07/21]

(3) You see the name – the printed name is "*Ekatome Rambo John*". Our client's name is *Alfred Rombhot Yekatom*. Would you agree it seems likely a person would not have signed that without indicating the changes? [Defence, 12/05/21]

(4)   PRESIDING JUDGE SCHMITT: *Rombhot* is a synonym, or could be, I'm not sure, could be – it could mean *Yekatom*. You said when you came to the ceremony, you learned that the children or that some of the children were from *Rombhot's* – and this might be *Yekatom*, you can tell us – *Yekatom's* group. [04/11/21]

Felton Rosulek (2015: 57) argues that an additional category of informal nominations using nicknames should be added to van Leeuwen's taxonomy to capture the meaning nuances when indexing different identity facets. In example (2), the use of a nickname portrays in-group solidarity with the defendant to which the witness refers. Different spellings of the nickname gave rise to some discussion as to the identity match (3) and, in general, were used cautiously by the Court (4) as a means of unequivocal identification. This was for the benefit of the extended

Table 2: Terms of reference used for the defendants in the whole corpus and in the opening statements.

Term of reference	Whole corpus (AF)	Range (texts)	Whole corpus (NF)	Prosecution OS (NF)	Victims' representatives OS (NF)	Ngaïssona's Defence OS (NF)
Ngaïssona	1,922	100%	171,312	**654**	11	**425**
Mr Ngaïssona	1,266	96%	112,841	41	3	**345**
Patrice-Edouard Ngaïssona	72	98%	6,418	3	2	2
Mr Patrice-Edouard Ngaïssona	8	13%	713	0	0	0
Patrice Ngaïssona	32	27%	2,852	0	0	**28**
Yekatom	611	100%	54,460	**435**	8	0
Mr Yekatom	312	98%	27,809	116	3	0
Mr Alfred Yekatom	3	4%	267	0	0	0
Alfred Yekatom	65	91%	5,794	27	2	0
Rombhot	63	96%	5,615	17	0	0
Alfred Rombhot Yekatom	46	96%	4,100	0	0	0
Rambo	58	24%	5,170	31	5	0
Defendant	9	16%	802	0	0	6
Defendants	1	2%	89	0	0	0

participation network, including witnesses and victims who were used to identifying the person in that way.

Remarkably, the single-standing functionalization "defendant", notoriously used by prosecution lawyers to background the identity of the persons involved, or to portray them in a negative light, stripping them of their humanity (see, e.g. Tiersma 1999: 188 on the use of "defendant" by the Prosecution in O.J. Simpson's case), was used only nine times and only by one defence counsel in relation to his client (5).

(5)   MR KNOOPS: Mr President, *the defendant* asks for assistance of the Registry. He has no sound on his – [05/07/21]

This peculiar choice might be interpreted as a desire to place additional distance between the defence counsel and his client, or simply be a sign of a more formal community of practice, as participants come from different national systems. The plural "defendants" was used only once by the Presiding Judge when giving instructions to the Prosecution aiming to improve time-efficiency of the proceedings. The Judge invited the Prosecution to focus their witness examination on the events involving the defendants. This hapax is placed against the general background of formal nominations ensuring respect for the identity of the people on trial.

While both the opening statements by the defence and the prosecution abound in defendants' nominations, they are largely underrepresented in the opening statements of the victims' representatives, whose story concentrates on the voices of the victims (see 4.3). In other words, by de-emphasizing the defendants and their stories, the victims' representatives index victims as more active and ratified participants in the proceedings.

4.2 Institutional parties

The Court was verbally represented as either "the Court" or "the Chambers". The latter term (6) was used in a highly formulaic gratitude formula which the Presiding Judge recited after every witness examination, with small variations, to index the sitting panel of judges specifically and the whole institution they represent. "The Court" as a vocative was used predominantly by other parties (7) to refer to the judges, and not the Presiding Judge who spoke for the judges.

(6)   Madam Witness, *on behalf of the Chamber* I would like to address you. *On behalf of the Chamber* I would like to thank you for coming, in these difficult times, to this far away country to give testimony here to assist *the Chamber* in establishing the truth. [Presiding Judge, 16/03/21]

(7)   *Mr President, your Honours*, my name is Claire Henderson and may it please *the Court*, I appear for the Prosecution this morning. [17/02/21]

The Presiding Judge, Judge Schmitt, who, with no exceptions, was the only ratified speaker for the Chambers, was addressed predominantly as "Mr President" (7), yet other speakers were careful to include the other two members of the chambers with the identification "Your Honours" (7); in fact, a left collocate search (L5) proved that these two titles were frequently used within the same phrase (7). Judge Schmitt was also referred to with the more generic identification "Your Honour", however, such generic identification was significantly less frequent than his main functionalization "Mr President" (see Table 3). The Chambers' members names were specified only once; this occurred during the very first hearing, after which they were referred to only by functionalization or identification, but never nomination, indexing their institutional impartiality.

Table 3: Terms of reference for the institutional participants.

Term of reference	Whole corpus (AF)	Whole corpus (NF)	Prosecution's OS (NF)	Victims' representatives' OS (NF)	Ngaïssona's Defence OS (NF)
The court	821	73,178	14	52	46
Mr President	1,602	142,790	68	19	**214**
Mr President * Your Honours	264	23,531	48	8	68
Your Honour	260	23,174	21	12	2
Your Honours	496	44,210	**359**	34	110
The Chamber	416	37,079	10	8	98
Mr Knoops	995	88,687	n/a	n/a	n/a
Mr Vanderpuye	798	71,127	n/a	n/a	n/a
Ms Dimitri	536	47,775	n/a	n/a	n/a
Mr Hannis	355	31,642	n/a	n/a	n/a

Table 3 (continued)

Term of reference	Whole corpus (AF)	Whole corpus (NF)	Prosecution's OS (NF)	Victims' representatives' OS (NF)	Ngaïssona's Defence OS (NF)
My colleague	43	3,833	62	5	2
Counsel	333	29,681	0	5	2
The Prosecution	479	42,694	62	22	**283**
The Defence	274	24,422	0	12	36
Prosecution	601	53,568	75	22	337
Defence	554	49,379	**3**	14	84
Team	146	13,013	10	5	0
Defence Team	23	2,050	0	0	0
Our defence team	6	535	0	0	22
Parties	157	13,994	0	5	2
Everyone in the courtroom	17	1,515	0	0	0
Everyone	279	24,868	17	2	0
Learned friend	15	1,337	0	0	2

Table 3 shows that the Prosecution ("Your Honours", NF=359) and the Defence ("Mr President" NF=214) appealed directly to the Court significantly more frequently in their respective opening statements than the representatives of the victims. In addition, the Defence referred to the Prosecution more often (NF=283) than the Prosecution referred to the Defence (NF=3). A very peculiar trend was observed in the speech of Judge Schmitt, who extended the ratified participants framework to "everyone in the courtroom" (8) on five occasions, starting with the first witness examination on 16/03/2021 and then additionally on 30/04/21, 26/05/21, 04/06/21 and 09/07/21. This pattern appeared in the speech of one of the defence counsel, Mr Knoops, on 04/06/21, who then used it on ten occasions altogether, mirroring and aligning with the judge's style (7). Coupled with the occasional vocative "dear members" (9), this discursive alignment may be interpreted as a clear attempt at creating a common ground.

(8)   It is also important to know, for example, for *everyone in the courtroom* and for the Chamber [. . .] [Presiding Judge, 09/07/21]

(9)   Good morning, Mr President, dear members of the Chamber, *everyone in the courtroom*. The Defence team of Mr Ngaïssona appears today before you [. . .] [Mr. Knoops, 13/07/21]

The parties' representatives were indexed with a functionalization "counsel", used both for defence and for prosecution counsel. For vocatives, they were always unequivocally addressed using the nomination of the type "Mr"/ "Ms" + surname (several examples are reported in Table 3), thus drawing a demarcation line between the judges, who were never addressed using nomination, and other participants.

The expression "team" or rather "Defence team", peculiarly, was used only by the team of Ngaïssona, who either referred to themselves as "the Defence team of Mr Ngaïssona" (7) or "our Defence team" (six occurrences), combining thus the functionalization with a personal pronoun. The use of the first-person plural pronoun, along with a collective noun *team*, establishes "an interactional group" (Wortham 1996: 322) and gives "disparate individuals a common identity" (Felton Rosulek 2015: 156). Interestingly, the analogical reference "the Prosecution team" was used only once, showing the dispreferred character of this term of reference. The standard appellative appears to be "the Prosecution", which is used almost twice as frequently as "the Defence".

The use of "we"-rhetoric for group identity was very different for the Defence and the Prosecution in their opening statements. In line with "our team" approach, the Defence used "we" (NF=423) and "our" (NF=232) to refer to themselves and to include other participants in the proceedings, in an attempt to create in-group solidarity (10).

(10)   Now, why are *we* telling this today to this Chamber? Why are *we* resuming the history of Mr Ngaïssona? Why is this relevant for this trial? (Defence, 18/02/21)

(11)   They were told that they were, quote, "going to fight the Arabs and that everybody had to be alert . . . there was no difference between the men, women and children and *we* had to kill them all." (Prosecution, 17/02/21)

At first glance, the first-person plural rhetoric is prominent in the Prosecution's statements ("we" NF=250), too. Yet, the analysis of concordances shows that most instances of "we" referred to out-group identity (11), excluding the Prosecution and the public, and referencing the perpetrators of the events that were depicted as an organized group.

4.3 Witnesses and victims

As many of the witnesses were testifying under protection, their identity was hidden, and they were consistently referred to using functionalization, "witness" (see Table 4), to which the honorific "Mr" or "Madam" was added at times (6). It can be hypothesized that the dichotomy between the pattern "Mr"/ "Madam" + "witness" and the standalone "witness" was introduced at the sanitization and transcript redaction stage, as the co-text analysis does not provide a clear explanation for this phenomenon. The witnesses were specifically addressed by this functionalization, frequently coupled with the second-person pronoun, unambiguously signalling that they are the ratified speakers during the next turn.

Table 4: Terms of reference for witnesses.

Term of reference	Whole corpus (AF)	Whole corpus (NF)	Range	Prosecution's OS (NF)	Victims' representatives' OS (NF)	Ngaïssona's Defence OS (NF)
Witness	6,068	540,854	100%	147	3	112
Witnesses	173	15,420	73%	82	3	98
Mr Witness	987	87,973	80%	0	0	0
Madam Witness	132	11,765	9%	0	0	0
Witness * you	160	14,261	n/a	0	0	0

During the proceedings analyzed, only six out of seventeen witnesses appeared in-person, with the majority attending online, exploring the digital opportunities of blended courtroom participation. The Presiding Judge recognized the Prosecution's difficulty in bringing witnesses in-person to The Hague, "under the circumstances, pandemic and also the circumstances in the country" (19/03/21, p. 2, line 17). The Judge ratified each witness's status by addressing them – either by name or by the role, as discussed above, for redacted cases – and framed the visible on-screen face[2] as a marker of a participant's status (1; 12). Even if nine out of eleven remote witnesses appeared under witness protection, i.e. their faces and voices were distorted, the camera always held their image in a close-up.

(12) Mr Demafouth, if you feel more comfortable, you may take off your mask. That may be perhaps – it's up to you, but I think we would prefer if you could – *if we could see you* and is perhaps also more comfortable for you. Okay. Again, good morning, *now we have a face in front of us*. [Presiding Judge, 05/07/21]

It was illustrated in the previous section how Judge Schmitt used "everyone in the courtroom" (8) as a reference term, echoed by the counsel who transformed it into a vocative (9) on most occasions, thus creating a group with boundaries defined by a physical presence in the courtroom. This may be interpreted as potentially excluding remote participants (witnesses on the dates in question) or extending the courtroom to what was referred to as "video-link location", see (13). The initial ritual interchange anchored every active participant as a ratified speaker to the reality of the proceedings and silenced the differences between in-person and remote participant status. This was even though some extra formulas aimed at verifying connection (13) were added (presence of simultaneous interpretation, possibility to view evidence on the shared screen when necessary):

(13) And we have of course also *the witness at the video-link location*. Good morning, Mr Witness. There seems *not to be a connection* at the moment. It should be fine. I ask you again, Mr Witness, good morning. *Do you hear me and understand me well?*

Schmitt (2021: 507) posited that "victims' participation under the Rome Statute is rightly considered a monumental development in international criminal proceedings and one of the cornerstones of the ICC Statute". However, whereas witnesses were central to the proceedings, victims occupied a peripheral position in terms of their status ratification. "Victim" as a term has been problematized (see Felton Rosulek 2015: 107–108) in that it can carry negative associations, yet in the case at

---

2 Face-to-face exchanges become particularly relevant with the integration of video technology into the proceedings. Video shots may either impinge on someone's face, thus somewhat deratifying the person of their participant's status or, vice versa, may secure or signal the active status by zooming in on someone's face. As the video materials were incomplete in this case, the analysis excludes the multimodal insights. For multimodality and the value of the face in remote courtrooms see, e.g. Matoesian and Gilbert (2018) and Licoppe (2021).

hand it was at times even intentionally claimed, as (14) illustrates. As the witness in (14) was an expert witness, his testimony was not redacted in contrast to many other cases, allowing here the assessment of the use of deictic "I" in anchoring his personality to the secondary reality of events in the CAR and to the primary reality of the proceedings. It was one of the rare cases where "victim" collocated with "I", i.e. with active speaker status.

(14) All what the country lived through, I do not know *who is a victim* and who is not. If you look into history in that context – now I am listening to you talk about Muslims and victims. It is difficult to accept. In any case, *I'm a victim*. I am a civilian servant. And the entire time I was organising myself to take care of my family, but all the property I accumulated was looted. *So I was a victim.* [Mr Ngaya, 04/11/21]

(15) The situation of *victims* before the International Criminal Court is paradoxical. They are not at the centre of investigations and the trial, but, rather, they find themselves on the periphery, overshadowed by one relationship which is in the spotlight, that is to say, the relationship between the Prosecution and the Defence in connection with the accused. This position of the *victims* is somewhat disadvantageous and, in fact, is a type of secondary *victimisation*. [Ms Rabesandratana, Legal Representatives of Victims, 17/02/21]

The plural "victims" was used almost ten times more frequently than the singular "victim", with a similar tendency concerning "children" or "child soldiers". The latter represented a particular category with a separate legal representative as the case concerned, inter alia, conscription of children in armed conflict. These plural forms indexed the preference to treat victims as a collective body of participants rather than individuals, which indeed may be interpreted as secondary victimization because it excluded individual voices (15).

Analysis of the concordances showed that the numbers are skewed, in that half of the mentions of "victim(s)" occurred during the opening statements of the common legal representatives of victims, the legal representative of former child soldiers and the Prosecution. In other words, victims' representatives and the Prosecution defined victims as more central to their narrative, which is not surprising.

The singular form "victim" tended to be preferred by few participants, notably the legal representative of former child soldiers, Mr. Suprun (16), who – on twelve occasions – recurred to the sanitized nomination, rather than a simple functionalization "victim", thus actually ratifying single people as participants. In a comparative perspective, the Prosecution used the singular "victim" on eight occasions (four by one lawyer and four by another) and only as functionalization, so showing a lower degree of personalization. The Defence team of Ngaïssona used "victim" and "one of victims" only twice to refer to the defendant himself, reversing the narrative and reclaiming the term. By not mentioning the victims, the Defence silenced their identity and their stories, focussing instead on Ngaïssona's personality (see Table 5).

(16) To illustrate, I'm going to cite the experiences of *certain former child soldiers*. We have *victim a/65131/19*, who was conscripted into the Anti-Balaka when he was only 12 years old, and he was obliged to attend summary executions of civilians. *Victim a/65122/19* was also enrolled at the age of 12 years and tells how that victim was obliged to watch and learn how to cut up people. [Mr Suprun, 17/02/21]

Table 5: Terms of reference for victims.

Term of reference	Whole corpus (AF)	Whole corpus (NF)	Prosecution's OS (NF)	Victims' representatives' OS (NF)	Ngaïssona's Defence OS (NF)
Victims	457	40,733	62	**255**	2
Victim	53	4,724	24	37	2
Child soldiers	120	10,696	24	45	0
Child soldier	9	802	7	2	0
Children	299	26,651	147	25	12
Muslims	588	52,410	**414**	29	20
Séléka	1,592	141,898	**431**	8	64
(Muslim) civilian population	92	8,200	72	6	0
Civilians	317	28,255	233	8	4

It comes as no surprise that most mentions of victims are made by the representatives of victims, where victims are portrayed as a homogeneous group and a quintessential "they" (NF=225 in the opening statements), which collocates predominantly with verbs expressing mental processes (Halliday 1994), such as "hope", "believe", "expect", "wish", etc, see (17).

(17) So the *victims rely on* the Court to shatter the culture of impunity, as well as the spiral of violence which engulfs the Central African Republic, which is a source of unhappiness, and *they hope* that there will be a deterrent decision by the Court, and *they have expressed* this by saying "enough is enough". (Legal Representative of Victims, video-link, 17/02/21)

Although their concerns are expressed and acknowledged, the sheer amount of victims in war crime proceedings excludes single voices, replacing them by a ratified and homogeneous "they". As Maître Fall (the victims' representative) said, participation of victims is realized only through a proxy of their representatives and even that "participation is very limited" (17/02/21, p. 59, line 15).

Frequently, the victims were referred to by the group identification "Muslims" (or less frequently "Arabs" in quotes), showing the ethnical-religious roots of the conflict, or "Séléka" (the rebel coalition), omitting their victim status. Alternatively, they were indexed as "civilians" or "civilian population" by the Prosecution, shifting the narrative to a more geopolitical representation.

4.4 Interpreters

The simultaneous interpreter's "invisibility" has received a lot of scholarly attention (Angelelli 2004; Berk-Seligson 2017; Ozolins 2016). To use Goffman's framework, interpreters take on the so-called "animator" role, almost a "talking machine", i.e. the body producing the sound, but not necessarily its "author" (Goffman 1981: 167). Building on Goffman, Wadensjö (1998: 92) analysed the interpreter's status and referred to it as "reporter" footing, i.e. saying the same as the original speaker but in a different language. Indeed, whenever a participant was speaking in a language different from English, the English transcript marked it with "interpretation" in parenthesis, indexing their reporter status.

In the present case, the interpreters were repeatedly acknowledged by ratified speakers in their "voiceover" / "linguistic bridge" function, as shown in (18) – (19).

(18) I'll begin with just a few words in French, *for the interpreters*. (Prosecution, 16/02/21)

(19) Our various examples – and *for the translation booth* I will now skip several examples in my submissions and I will go to paragraph 37 of my presentation (Ngaïssona Defence Opening Statement, 18/02/21)

The examples above show that, despite the interpreters being an integral part of international proceedings, there is still some variability as concerns the labels attributed to them (see Table 6), including also the metonymic "booth".

Table 6: Terms of reference for interpreters in the whole corpus.

Term of reference	Absolute frequency	Normalized frequency	Range (texts)
Interpreters	126	11,231	76%
Interpreter	294	26,205	89%
Translators	4	357	4%
Translation	83	7,398	62%
* booth	35	3,120	27%

However, there were multiple cases, when interpreters were ratified as "responders" combining the functions of "animator", "author" and "principal" (Wadensjö 1998: 92) and taking personal responsibility for the utterance (Wadensjö 1998: 165). This type of participation can be illustrated by instances of repair (20) and clarification (21).

(20) MS DIMITRI: I would like to show you a photograph, CAR-OTP-2122–9339, tab 13. It is not to be broadcast in public.
THE INTERPRETER: *Correction: It ends with 2399.* (16/03/21)

(21) THE INTERPRETER: Inaudible. *The interpreter didn't hear* the last part of the witness's answer.
PRESIDING JUDGE SCHMITT: Mr Witness, the interpreter and also we, of course, did not get the second part of your answer. Would you please be so kind to repeat it. (11/05/21 – remote witness via interpretation)

In examples (20) and (21), interpreters became ratified speakers; however, their statements were either impersonal (20) or relied on third-person deixis (21). Whenever the interpreters went beyond mere reporting, their remarks were marked in the transcript as "the interpreter", and, as Table 6 illustrates, these were both frequent and dispersed across multiple hearings. Interestingly, there were several occasions where interpreters intervened as active ratified speakers or "principals" (Goffman 1981: 226), i.e. the ones whose opinion was uttered and who were bound to what was being said (Goffman 1981: 144). This was also linguistically marked through the ratified-speaker first-person deixis, as exemplified in (22) below:

(22) MS DIMITRI: I'm sorry, Mr President, I'm listening in French, there's no – there's *no French translation*. There's an issue. There hasn't been *any French translation* for the last exchange you had with –
THE FRENCH INTERPRETER: Excuse *me*, sir. This is *the French booth*. This is the chef d'équipe. The problem is that *we are sharing* one French link with the Sango booth, Sango into French, English into French, and the Sango booth into French had taken up the French channel so *I* could not have access to the French channel. This is going to have to be solved.
PRESIDING JUDGE SCHMITT: But actually, when I hear that, it seems to be an issue that can be solved.
THE FRENCH INTERPRETER: Yes, it's an issue that can be solved. *We* have to work it out between *the two booths, the Sango-French booth and the English-French booth*.
PRESIDING JUDGE SCHMITT: No problem.
THE FRENCH INTERPRETER: Okay, sorry.
PRESIDING JUDGE SCHMITT: No, no, you don't have to be sorry. These are, you know, at the beginning (Overlapping speakers)
THE FRENCH INTERPRETER: Teething problems, yes. [. . .]
Maybe it would be best, President, *this is the French booth* again, it would be best maybe to have just a short break so *we* can liaise and so *I* can ask the AV if there is a problem because it seems that there is also an AV problem, to take the floor.

Example (22) clearly illustrates how the interpreter took on the role of an active ratified speaker, explaining the technical issue to the Presiding Judge and clearly taking on responsibility for the content of the utterance. Although such personal exchanges were not frequent in the corpus, their very presence confirms that interpreters in international courts are active participants with varying degrees of status ratification.

5 Discussion and conclusions

This study stemmed from the desire to add a linguistic perspective to what can count as effective participation in a digitalized context of an international court. By creating groups, and anchoring different aspects of identity to the reality of proceedings, participants are able to foreground certain elements and silence others, leading to the creation of different stories. Many participants had to join the proceedings online, leading to increased digitalization but questioning the effectiveness of their participation, which had to be ensured through new courtroom rituals. Like an orchestra director, the Presiding Judge at the ICC conducted the hearings, curbing any participation differences that could have arisen on account of different participation modes – remote or in-person, ensuring both the discursive and the visual presence of the ratified participants. Most participants were referenced using nomination, with the exception of the members of the Chambers, who introduced themselves by name only once, and the witnesses under protection, whose personal details were undisclosed. It appears that nomination is equalled to the Court's linguistic realization of effective participation. A slight difference emerged as to the use of titles "Mr" with the defendants' surnames. The Prosecution more significantly used the defendants' surnames, unaccompanied by any honorifics, thus de-emphasizing the social dimension of their personality, whereas the Defence frequently used the defendant's first name in the formula "Mr" + name + surname, thus portraying a more humane picture of the man and confirming previous studies on defendants' (de)humanization (Tiersma 1999).

While there was no significant divergence in the reference terms used for the main institutional participants, the frequency of their use may suggest an attempt at in-group identity creation by the Defence team, who frequently invoked the Prosecution (4.5 times more often) and the Presiding Judge (3 times more often), as well as engaging in the inclusive "we"-rhetoric, thus placing themselves on the same procedural, if not discoursal, level. The Prosecution did not single out the Presiding Judge, preferring the general categorization vocative "Your Honours" (10 times more often than Victim's representatives and 3 times more often than the Defence).

Different participants emphasized different parts of the narratives and ratified different actors as relevant for their stories, which became especially prominent in the case of victims. Although the ICC is declaredly proud of including the victims among the trial participants, their participation remains limited, not only factually but also linguistically. Victims were ratified as unaddressed listeners by the Prosecution, who included their identification with general categories (Muslims, Sélékas) in the narratives of the past events which were the basis for the proceedings. The Defence (of Ngaïssona in this case) silenced victims' narratives altogether reclaiming the title for the defendant, who was presented as a victim himself. Only the legal representatives of victims referenced them in the present as the agents of mental processes, making requests to the Chambers, with future projections.

Finally, the findings contributed to the ongoing debunking of the myth concerning the interpreter's invisibility, showing that in international trials interpreters are not only ratified as reporters, but are also given an opportunity to speak for themselves, which makes it possible to conclude that simultaneous interpreters at the ICC may be counted as ratified speakers.

References

Angelelli, Claudia. 2004. *Revisiting the interpreter's role: A study of conference, court, and medical interpreters in Canada, Mexico, and the United States*. Cambridge: Cambridge University Press.

Baker, Paul, Costas Gabrielatos, Majid Khosravinik, Michał Krzyżanowski, Tony McEnery & Ruth Wodak. 2008. A useful methodological synergy? Combining critical discourse analysis and corpus linguistics to examine discourses of refugees and asylum seekers in the UK press. *Discourse and Society* 19(3). 273–306.

Berk-Seligson, Susan. 2017. *The bilingual courtroom: court interpreters in the judicial process* (2nd edn with a new preface). Chicago: University of Chicago Press.

Biber, Douglas, Stig Johansson, Geoffrey Leech, Susan Conrad & Edward Finegan. 1999. *Longman grammar of spoken and written English*. Essex, UK: Pearson Education.

Drew, Paul &John Heritage. 1992. Analyzing talk at work: An introduction. In Paul Drew & John Heritage (eds.), *Talk at work. Interaction in institutional settings*, 3–65. Cambridge: Cambridge University Press.

Eckert, Penelope & Sally McConnell-Ginet. 1992. Think practically and look locally: Language and gender as community-based practice. *Annual Review of Anthropology* 21, 461–490.

Council of Europe. 1950. *Convention for the protection of human rights and fundamental freedoms*. (Council of Europe Treaty Series 005), Council of Europe.

Felton Rosulek, Laura. 2015. *Dueling discourses: The construction of reality in closing arguments*. Oxford: Oxford University Press.

Goffman, Erving. 1967. *Interaction ritual. Essays on face to face behavior*. New York: Anchor Books.

Goffman, Erving. 1974. *Frame analysis. An essay on the organization of experience*. New York: Harper & Row.

Goffman, Erving. 1981. *Forms of talk*. Oxford: Blackwell.

Halliday, Michael A. K. 1994. *An introduction to functional grammar*. Baltimore, MD: Edward Arnold.

ICC Case information sheet. 2021. ICC-PIDS-CIS-CARII-03-012/20_Eng (Updated: July 2021). Available at https://www.icc-cpi.int/sites/default/files/2022-04/yekatom-ngaissonaEn.pdf (Accessed on 30 June 2022).

ICC Policy paper on case selection and prioritisation. 2016. ICC Office of the Prosecutor (15 September 2016), available at www.icc-cpi.int/itemsDocuments/20160915_OTPPolicy_Case-Selection_Eng.pdf (accessed on 30 June 2022).

International Commission of Jurists. 2020. *Videoconferencing, Courts and COVID-19 recommendations based on international standards*. Available at https://www.unodc.org/res/ji/import/guide/icj_videoconferencing/icj_videoconferencing.pdf (Accessed on 30 June 2022).

Leeuwen, Theo van. 2002. The representation of social actors. In Michael Toolan (ed.), *Discourse analysis*, 302–339. London: Routledge.

Licoppe, Christian. 2021. The politics of visuality and talk in French courtroom proceedings with video links and remote participants. *Journal of Pragmatics* 178, 363–377.

Matoesian, Gregory & Gilbert, Kristin. 2018. *Multimodal conduct in the law. Language, gesture and materiality in legal interaction*. Cambridge: Cambridge University Press.

Mikkelson, Holly. 2016. *Introduction to court interpreting*. 2nd edn. London: Routledge.

Morford, Janet. 1997. Social indexicality in French pronominal address. *Journal of Linguistic Anthropology* 7(1), 3–37.

O'Barr, William. 1982. *Linguistic evidence: Language, power and strategy in the courtroom*. New York: Academic Press.

Ozolins, Uldis. 2016. The myth of the myth of invisibility? *Interpreting* 18(2). 273–284.

Partington, Alan. 2004, Introduction: Corpora and discourse, a most congruous beast. In Alan Partington, John Morley & Louann Haarman (eds.), *Corpora and Discourse*, 11–20. Frankfurt am Main: Peter Lang.

Rome Statute of the International Criminal Court (last amended 2010), adopted on 17 July 1998 by the United Nations Diplomatic Conference of Plenipotentiaries on the Establishment of an International Criminal Court, entered into force on 1 July 2002. Available at https://www.icc-cpi.int/sites/default/files/RS-Eng.pdf (Accessed on 30 June 2022).

Rossner, Meredith & Tait, David. 2021. Presence and participation in a virtual court. *Criminology & Criminal Justice*, 174889582110173–. https://doi.org/10.1177/17488958211017372.

Sacks, Harvey, Emanuel A. Schegloff & Gail Jefferson. 1974. A simplest systematics for the organization of turn-taking for conversation. *Language* 50(4). 696–735. https://doi.org/10.2307/412243.

Salaets, Heidi,Katalin Balogh & Dominique Van Schoor. 2016. Languages of lesser diffusion (LLDs): the rationale behind the research project and definitions. In Katalin Balogh, Heidi Salaerts & Dominique Van Schoor (eds.), *TraiLLD: Training in languages of lesser diffusion*, 17–34. Leuven: Lannoo Campus Publishers.

Schmitt, Bertram. 2021. Legal diversity at the International Criminal Court. *Journal of International Criminal Justice* 19(3). 485–510. https://doi.org/10.1093/jicj/mqab038.

Swigart, Leigh. 2019. Unseen and unsung: ICC language services and their impact on institutional legitimacy. In Freya Baetens (ed.), *Legitimacy of unseen actors in international adjudication*, 1–28. Cambridge: Cambridge University Press.

Tiersma, Peter. 1999. *Legal language*. Chicago: University of Chicago Press.

United Nations. 1948. *Universal declaration of human rights*. New York: United Nations General Assembly.

Wadensjö, Cecilia. 1998. *Interpreting as interaction*. London/New York: Longman.

Wortham, Stanton. 1996. Mapping participant deictics: A technique for discovering speakers' footing. *Journal of Pragmatics* 25(3). 331–348.

Katia Peruzzo & Federica Scarpa

Making national immigration and asylum case law accessible to a non-Italian audience: The role of intra-lingual translation

1 Introduction

In European Union (EU) Member States, international protection is governed by a number of international, European and national rules, where a key role is played by European Union law. Ever since the 1951 Geneva Convention on the protection of refugees, asylum has been considered a fundamental right and an international obligation for the Convention's signatory countries. This has led the European Union to develop a body of legislation to ensure minimum standards of protection to third-country nationals or stateless persons in the form of asylum or subsidiary protection. Based on the Geneva Convention, the Amsterdam Treaty in 1997 introduced the idea of an "Area of Freedom, Security and Justice". This was followed by an extraordinary European Council meeting at the Finnish city of Tampere in 1999 where a "Common European Asylum System" (CEAS) was established in order to create a genuine European Area of Freedom, Security and Justice where "individuals and businesses should not be prevented or discouraged from exercising their rights by the incompatibility or complexity of legal and administrative systems in the Member States" (European Parliament 1999). After laying down, in the following years, only "minimum rules" to harmonise reception conditions for asylum seekers and migrants in different States, CEAS was recast with the Treaty of Lisbon, which entered into force in December 2009 and made the adoption of "a common policy on asylum, subsidiary protection and temporary protection" possible by the EU. By developing a body of legislation to ensure minimum standards of protection to third-country nationals or stateless persons,[1] EU Member States have thus adopted a joint

---

1 The most important legislative acts aimed at reforming CEAS were the following:
1. Directive 2011/95/EU of 13 December 2011 on standards for the qualification of third-country nationals or stateless persons as beneficiaries of international protection, establishing two forms of international protection, i.e. the refugee status and subsidiary protection;
2. Directive 2013/32/EU of 26 June 2013 on common procedures for granting and withdrawing international protection;
3. Directive 2013/33/EU of 26 June 2013 laying down standards for the reception of applicants for international protection;

https://doi.org/10.1515/9783111048789-008

approach to welcome people fleeing persecution or serious harm in their country of origin, which they implement through national administrative and judicial authorities. The Treaty of Lisbon also determined an extension of the jurisdiction of the European Court of Justice and the entering into force, in 2009, of the Charter of Fundamental Rights of the EU, containing a number of significant rules concerning international protection, such as the prohibition of torture and inhuman or degrading treatment or punishment (Art. 4), the right to asylum (Art. 18) and the right to an effective remedy and to a fair trial (Art. 47), which has taken a fundamental role in the jurisprudence of the Court on asylum.

This paper focuses on the Italian case law developed through disputes concerning the rejection of applications for asylum or subsidiary protection by the Court of Cassation, i.e. the highest judicial authority in Italy. More specifically, it presents TrIACLE (Translated Immigration & Asylum Case Law in Europe), a collaborative translation project conducted at the Università degli Studi di Trieste whose underlying idea is that the monolingual tradition of Italian courts constitutes a limit to the dissemination of legal knowledge and the circulation of 'food for thought' beyond national boundaries. This lack of communication is all the more regretful given that Italian judicial decisions related to asylum and international protection contribute to a "common European law" on these crucial issues. By translating a selection of the Italian Court of Cassation's most recent and significant decisions on immigration and asylum into English, and by disseminating such translations, the project's main aim is to make the best practices of the Italian jurisprudence available to a wider audience. This audience consists of non-Italian-speaking international legal practitioners from other EU Member States whose knowledge of the Italian judicial decisions would otherwise be limited to problematic cases adjudicated by the European Court of Justice.

The first part of the paper by Federica Scarpa (sections 1–4) describes: 1) the interdisciplinary translation team in charge of the project (involving both specialists of EU law, linguists specialised in legal translation and MA students of Specialised Translation); and 2) the methodological choices and activities to overcome the challenges posed by both the culture-boundedness of the legal system underlying the source texts and the peculiarities of Italian judicial drafting. The second part of the paper by Katia Peruzzo (sections 5–6) provides concrete examples to show the

---

4. Regulation (EU) No 604/2013 of 26 June 2013 establishing the criteria and mechanisms for determining the Member State responsible for examining an application for international protection lodged in one of the Member States by a third-country national or a stateless person (the so-called 'Dublin system') and the so-called 'EURODAC' Regulation for the comparison of fingerprints for the effective application of the Dublin system.

fundamental role of the intra-lingual translation to make national case law intelligible to the international, and not clearly defined, audience of the translations.

2 The TrIACLE team

The project was initiated by the chair of EU Law at the Department of Law of the Università degli Studi di Firenze, assisted by Luca Minniti, judge at the Florence Court of Justice (Division of Immigration and International protection), in partnership with the two chairs of English Language/Translation and EU/International Law, both at the Department of Law, Language, Interpreting and Translation (IUSLIT) of the Università degli Studi di Trieste. In 2020–2021, IUSLIT conducted a first implementation phase of the project, called TrIACLE (Translated Immigration & Asylum Case Law in Europe), and created, within IUSLIT, an interdisciplinary team of lawyers (Stefano Amadeo and Fabio Spitaleri) and linguists specialised in legal translation (Federica Scarpa and Katia Peruzzo). This team then coordinated a group of five students who were completing the MA in Specialised Translation at IUSLIT and who carried out the translation work as part of their final dissertations.

Each MA student translated 2–3 decisions (judgments or orders) of the Italian Court of Cassation into English, thus contributing to the development of a bilingual parallel corpus made up of 14 original Italian texts and their English translations. Whilst the MA students' lack of legal knowledge could be overcome by the expertise of the other members of the team, the fact that the students were asked to translate into English, their second language (L2), and not into Italian, their first language (L1), seems at first glance to be less easily justifiable, given that many professional associations in their codes of ethics urge members to work exclusively into their mother tongue (e.g. AITI 2013). However, the motivations for this methodological choice are both practical and theoretical, and are the same as in a similar previous project in which the two authors of this paper were involved, i.e. the translation into English of the Italian *Codice di procedura penale* (Code of Criminal Procedure) (Gialuz, Lupària, and Scarpa 2014, 2017), also involving an interdisciplinary team of lawyers and linguists (Scarpa, Peruzzo, and Pontrandolfo 2017). By observing the reality of translation in countries where "languages of limited diffusion" are used (e.g. Croatia, Denmark, Finland, Slovenia) and also where "major" languages such as Spanish and Italian are spoken, it is a fact that L2 translation is a regular practice owing to the increasing hegemony of English as the language of globalisation (Cronin 2003: 144–146) and – albeit unofficially – the language of the law at an international level. And it is precisely because of the special position of English as a *lingua franca* of the globalised world that traditional axioms on the directionality in translation practice (from L2 to L1) have been challenged not only by everyday translation practice, but also by some studies based on empirical research (see 'inverse translation' in Pokorn 2005, 2009).

3 Challenges and reference documentation

Given TrIACLE's aim to create a corpus of decisions and to make it available to a wider international audience, the first major challenge for the translators on the project was that the applicable national legislation in the cases under examination is in Italian. This legal-system boundedness, manifesting itself mostly – but not exclusively – at the levels of terminology and phraseology, ruled out the choice for the target texts of a national variety of legal English, i.e. embedded in a specific national legal culture and being used only by native speakers. Following the model previously adopted for the translation of the *Codice*, the choice of the target language has fallen instead on European English. The different formal characteristics of this international variety of legal English are linked to different underlying legal concepts which can either be derived from national legal concepts – amongst which also those relating to the Italian judicial system – or developed *ex novo* at the EU supranational level. However, as in the *Codice* project, also in TrIACLE "European English" has a wider meaning than that of only "EU English", i.e. the variety of English that over the past couple of decades has become the *de facto* lingua franca in the supranational legal setting of the European Union. Here, besides including EU English, European English also comprises different types of legal English manifesting themselves in Europe under different circumstances.

So, besides using official EU legislative texts such as directives, regulations and decisions (https://eur-lex.europa.eu/homepage.html) as key references, as well as the decisions of the Court of Justice of the EU (https://curia.europa.eu/jcms/jcms/j_6/en/) and the IATE termbase (https://iate.europa.eu), during translation, the project translators also used the following documentation as a reference, which was compiled into an extensive specialized corpus:

– Council of Europe documents, especially judgments of the European Court of Human Rights (ECtHR) involving decisions refusing applications for international protection in Italy (https://www.echr.coe.int/documents/cp_italy_eng.pdf);
– texts by the UN Refugee Agency (e.g. https://www.refworld.org/pdfid/53b676aa4.pdf), independent think tanks and forums for debate on EU affairs such as CEPS (Centre for European Policy Studies, www.ceps.eu), and research

centres on international migration, asylum and mobility such as the Migration Policy Centre (https://blogs.eui.eu/migrationpolicycentre/twists-and-turns-asylum-laws-italy);
- (occasionally) English translations of summaries of judgments by both the Italian Court of Cassation (e.g. https://www.asylumlawdatabase.eu/en) and the Italian Constitutional Court (https://curia.europa.eu/jcms/jcms/p1_2170123/en/ and https://www.cortecostituzionale.it/actionJudgment.do);
- (very occasionally) judgments by the Supreme Court of the United Kingdom (https://www.supremecourt.uk/).

Although the standards applied to the Italian Court of Cassation's decisions rely on EU and international law available in English, the second major challenge posed by the translation of the decisions was the peculiarities of the language used by Italian judges, whose often convoluted nature has been noted by many scholars, both in Italian Studies (among others, Cortelazzo 1995; Ondelli 2014) and Translation Studies (among others, Rega 1997; Scarpa and Riley 1999). These peculiarities are at both the morphosyntactic level of the texts, the most noteworthy being the use of very long and complex sentences with missing inter-sentential connectives signalling the logical relations between different segments of the text, and at the lexical level, with the possible exception of terminology, which instead is standardised across the 14 decisions. An example of a lexical trait which provided a challenge for the translators is the use of different variants to denote the same concept within the same decision: in Order No. 17954/2020, the Turin Court of Appeal (*Corte d'Appello di Torino*) is in fact referred to by as many as seven different terms (*Corte cisalpina, Corte distrettuale, Collegio subalpino, Collegio piemontese, Collegio cisalpino, Corte subalpina* and *Corte territoriale*), in compliance with the low level of tolerance to lexical repetition typical of the Italian rhetorical style, opting instead for semantically ambiguous synonymic variation (cf. Scarpa 2020: 141, 256–257). However, these peculiarities of Italian judicial drafting were not homogeneous throughout all the decisions which, although produced by the same institution (the Court of Cassation), were in fact drafted by different judges, each possessing their own individual style. Thus, for example, the subject-verb inversion and the frequent use of the *passato remoto* tense, which both typify Italian judicial texts, are particularly frequent in one decision (Order No. 17954/2020) but are totally lacking in another (Order n. 18541/2019).

4 Methodology

The first methodological choice to be made by the interdisciplinary team was closely linked to the aim of the TriACLE project, i.e. to make the Court of Cassation's decisions available to a wider audience of readers who do not speak Italian. Therefore, the choice concerned the general approach to be followed during translation – an approach which would guide the translators' decisions at all levels of the text. The predominantly informative function of the resulting translations, i.e. enabling non-Italian speakers to study the characteristics of the Italian legal system and language, entailed a variation of both use and addressees of the target texts as compared to those of the source texts, whose function was instead mainly performative. The resulting translations would in fact be neither "for normative purposes", i.e. would not have the legal force of the source text, nor "for general legal or judicial purposes", i.e. would not be used in court proceedings as part of documentary evidence, but should instead be "documentary translations" (Šarčević 1997: 19; Cao 2007: 10–12). The choice of this general approach meant that translators should use explicitation and simplification strategies as frequently as possible in order to make their target text clearer and more legible than the corresponding source texts. This requirement seems at first glance to contravene the vagueness and indeterminacy that characterise legal language (see Bhatia et al. 2005; Joseph 1995; Engberg and Heller 2008), which has to be maximally determinate and precise and, at the same time, all-inclusive, in order to cover every relevant situation. However, as Bhatia and colleagues (2005: 9–10) have remarked, in order to carry out their task effectively, it is part and parcel of legal translators' jobs to place strong emphasis on the exact phrasing of texts rather than on their vagueness, thus limiting the possible interpretation of the legal text. The variation of use of the translations also had consequences for the degree of completeness of the information contained in the target texts, as opposed to that in the source texts (cf. Scarpa 2020: 199–200), because it meant that the translators were asked to translate not whole decisions but only those selected portions providing the specific information relevant to the target-text users' purposes. For example, the initial page of the decisions, containing information on the Court composition and on the parties involved, was systematically eliminated from the texts to be translated, whose final length was consequently, on average, at least half of that of the original decisions.

A second methodological choice involved the work processes to be followed in order to, on the one hand, harmonise the five translators' choices concerning terminology/phraseology and style, and, on the other, maximise the contribution of the legal experts on the team.

The most innovative work procedure adopted in the translation project was the novel type of collaboration between the lawyers and the linguists on the team, which was developed only in the final stages of the project as it had not initially been planned. Indeed, the 14 decisions were divided into three lots and each lot underwent a pre-translation phase before being handed to the translators. The original idea was to follow the work procedure suggested by Gile (1986), where the specialist's competence comes into play not only *after* the text has been translated (as in the intra-lingual "specialist review" recommended by international translation standards such as ISO 17100: 2017) but, crucially, also *before* the beginning of the translation activity. According to Gile's translation procedure, before the actual translation stage the lawyers should identify any translation problems in the source texts (difficult terminology/phraseology, complex sentences etc.) and give the translators a list of all these problems, as well as information on how to solve them. However, possibly owing to the lawyers' lack of experience in the specific fields of translation and revision, the best formula for the project's type of team work – integrating specialist-domain competence with the translator's drafting and methodological competences – turned out to be the one adopted only for the third lot of translations, where the translation-oriented intra-lingual pre-editing of the source texts was carried out by both the linguists and the lawyers coordinating the project.

Concerning the organization and management of the project, the translation work was divided into three main phases, where the two linguists took on the role of joint project managers of the team:

– *Pre-translation phase*: An intra-lingual translation (pre-editing) of the source texts to be translated inter-lingually was carried out by both the lawyers and linguists on the team. The aim was to speed up the subsequent translation phase and increase the quality of its output by identifying and solving, beforehand, the problems the translators would encounter during their activity, as well as making the source texts more accessible and comprehensible. This was achieved by: 1) selecting the portions of the decisions which were going to be translated; 2) explicitating and simplifying any unnecessarily convoluted and complex (or even ungrammatical) syntactic structures; and 3) identifying problematic terminology/phraseology and suggesting possible equivalents. Whilst the identification and selection of the sections to be translated was carried out exclusively by the lawyers, mainly based on the two parameters of relevance to the new international readership and degree of innovativeness of the legal knowledge they contained, the other aspects of the pre-editing task were shared among lawyers and linguists. The division of work, however, was neither systematic nor consistent throughout the three lots into which the 14 decisions had been divided and handed to the five translators between March and May 2021. For the first lot, the texts were pre-edited by the lawyers only, whilst the second lot was pre-edited, first by the lawyers, this time more thoroughly than previously, and then by the linguists, who added their own comments and suggestions for possible solutions. The best practice for sharing the pre-editing work was however found only for the third and final lot. After the initial selection by the lawyers of the sections to be translated, the translation-oriented pre-editing was carried out only by the linguists, whose comments and questions to lawyers concerned the more problematic segments and terms of the texts, and often suggested possible solutions. In the course of subsequent meetings between the four linguists and lawyers, all the issues were discussed together, and possible solutions were added as comments in the text to be handed to translators. Both questions and answers/explanations were left visible in the texts for the translators.

– *Production phase*: The actual reformulation of the Italian source texts into English was carried out by the five translators both autonomously and collaboratively, given that the problems facing them were very similar in the 14 decisions, and the translations would all become part of the same final corpus. Thus, during their translation work, translators were constantly in contact with the other team members to solve problems and harmonise their translation choices. Being engaged in teamwork had the further advantage of reflecting real-life commissions where professional translators work in a team on the same project to increase the speed and quality of the actual translation work, including the effectiveness of the search for the appropriate terminology. In this phase, translators also identified and suggested solutions (duly quoting their sources) for the remaining terminological/phraseological problems which had not been identified and solved in the first pre-editing phase.

– *Revision phase*: The re-reading of the translations, to check how accurately the information contents and terminology of the source texts had been reformulated in the target language, was done in three stages, each following one of two different revision models reflecting the practices of professional translation teams, where the reviser is either a fellow translator (peer revision) or a senior translator ("pyramidal revision") (Pym 2015; Lafeber 2018). First, the translators revised each other's work according to the "translate-swap-revise" (or "mutual revision") model, where all the translators of a team are considered on an equal footing and systematically revise each other's work according to availability. Then the translations were revised by the two linguists on the project, jointly acting as senior translators leading the team of translators and deciding on the final translation to be delivered. The terminological/phraseological problems that were identified in this second revision were then solved at a number of meetings that the linguists had with the lawyers and the appropriate changes were then implemented by the translators. These restricted meetings were also preferred over a different task which, following the model adopted for the translation of the *Codice di procedura penale*, the team had agreed upon at the outset of the project, i.e. organising regular meetings of all project components, including translators, to discuss problems with the legal experts. These wider meetings were in fact organised at the very beginning of the project but did not prove to be successful for at least two reasons. The first was the relatively large number of TrIACLE translators, and their relative professional inexperience, as compared to the two, much more experienced translators of the *Codice*. The second was the medium of those meetings, which had to be held online because of COVID-19 pandemic restrictions, rather than face-to-face as in the past. In the third and last stage, the translators carried out a final, mutual revision to eliminate any remaining traces of inconsistencies left over from the previous rounds of revision of the 14 translations.

5 Pre-translation phase

Over two decades ago, Scarpa and Riley (1999: 56) argued that the language of Italian judgments was incomprehensible to non-lawyers. At the time of writing, the situation seems basically unchanged. For this reason, and in light of previous experience with translating Italian legal texts into English, at the project's outset the team decided that a pre-editing phase was necessary. The purpose of this was to enhance the readability and accessibility of the source texts, and thus increase the chances of obtaining a high-quality final product from the translation of the pre-edited texts rather than the original texts. As mentioned in the previous section, the pre-translation phase involved three main tasks, namely excluding the portions considered to be irrelevant for translation, explicitating and simplifying the source texts' morphosyntactic structures, and identifying problematic terminology and phraseology while suggesting possible suitable equivalents. Since the first task – carried out solely by the lawyers – only led to the signalling of omissions in the pre-edited texts (through the insertion of *[OMISSIS]*), in this section the focus is placed on the consequences of the other two tasks involved in the pre-translation phase, which were carried out jointly by both lawyers and linguists.

5.1 Morphosyntax

The intra-lingual translation led to a number of morphosyntactic changes in the source texts, the most common of which are illustrated below through concrete examples extracted from the judicial decisions under examination. The most visible changes are the following: i) the transformation of long sentences into short(er) sentences; ii) the inversion of deviant – though common in judicial language – verb-subject sequences into more common subject-verb sequences; iii) the transformation of impersonal or passive structures into personal and active structures; and iv) the conversion of nominal constructs into verbal ones.

5.1.1 Long sentences → short(er) sentences

A typical feature of Italian judicial decisions are long, convoluted sentences characterised by a high degree of hypotaxis and numerous incidentals and appositive phrases, all of which contribute to the complexity and low accessibility of these texts. In order to facilitate the production of the target texts, in the pre-translation phase such sentences were split into shorter sentences with fewer, or no, subordinate clauses, as in Example 1 (from Judgment No. 538/2019) below.

(1) Original version

È difatti richiamato a sproposito il principio secondo cui il diritto al riconoscimento dello status di rifugiato politico (o della misura più gradata della protezione sussidiaria) non può essere escluso, nel nostro ordinamento, in virtù della ragionevole possibilità del richiedente di trasferirsi in altra zona del territorio del paese d'origine ove non abbia fondati motivi di temere di essere perseguitato (o non corra rischi effettivi di subire danni gravi), atteso che tale esclusione, **prevista nell'articolo 8 della direttiva 2004/83/CE** ed il cui inserimento nell'atto normativo interno di attuazione della direttiva stessa costituisce una mera facoltà degli Stati membri, non è stata trasposta nel d.lgs. n. 251 del 2007 (Cass. 9 aprile 2014, n. 8399, in motivazione).

Pre-edited version

Il ricorrente ha affermato a sproposito che nel nostro ordinamento l'asilo (o la protezione sussidiaria) non possono essere esclusi quando il richiedente aveva la ragionevole possibilità di trasferirsi in un'altra zona del territorio del paese d'origine, nella quale non vi è il rischio che egli sia perseguitato (o subisca danni gravi). La possibilità di escludere l'asilo (o la protezione sussidiaria) per questa ragione **è prevista nell'articolo 8 della direttiva 2004/83/CE**. Si tratta di una facoltà degli Stati membri. L'Italia ha deciso di non avvalersi di questa facoltà. Questa possibilità di esclusione della protezione internazionale non è stata trasposta nel d.lgs. n. 251 del 2007 (Cass. 9 aprile 2014, n. 8399, in motivazione).[2]

Example 1 shows the original version (OV) and the pre-edited version (PEV) in parallel. The OV consists in a single, syntactically rather complex sentence of 116 words, with several subordinate and embedded clauses as well as parenthetical information provided in parentheses. The manipulation of the OV led to the transformation of a one-sentence paragraph into five separate sentences, for a total of 112 words. This change required a shift from a highly hypotactic composition to a sequence of shorter paratactic sentences with an occasional change in the distribution of information. For example, the OV presents the reference to the relevant legal system *nel nostro ordinamento* as an incidental phrase after the verb *non può essere escluso* and before the causal phrase. In the PEV, on the contrary, it precedes the subject *l'asilo*. What can also be noticed is that the passive structure at the beginning of the paragraph (*È difatti richiamato a sproposito il principio secondo cui*) was turned into an active form, which also entailed the explicitation of the agent (*Il ricorrente ha affermato a sproposito che*). This in turn led to a shift from a more nominal style to a more verbal style by deleting the noun *principio* from *è richiamato il principio secondo cui il diritto non può essere escluso*.

Another typical syntactic feature of Italian judicial decisions that contributes to their complexity, or even downright obscurity, is the high frequency of participial clauses and gerunds. In Example 1, an instance of a past participial clause is found in *tale esclusione prevista nell'articolo 8 della direttiva 2004/83/CE*, which in the PEV was not only transformed into a sentence of its own with a finite verb form, but also made even more explicit through the repetition of *asilo* and *protezione sussidiaria: La possibilità di escludere l'asilo (o la protezione sussidiaria) per questa ragione è prevista nell'articolo 8 della direttiva 2004/83/CE*. Despite being much shorter than the OV in Example 1, the OV in Example 2 (from Judgment No. 538/2019) contains two gerunds, which may be a source of ambiguity given that the subject is not always clear or explicitly stated. In this case, the first gerund (*censurando*) refers to the ground of appeal (*motivo*) while the second one (*non utilizzando*) is ideally relatable to the Court of appeal, which is however not mentioned in the text. Therefore, in the pre-translation phase both gerunds were excluded from the PEV, which led to the creation of three separate sentences and the explicitation of the relevant subjects.

---

2 'The appellant erroneously stated that, according to Italian national law, asylum (or subsidiary protection) may be granted even when the applicant could have reasonably moved to a different area of his country of origin where he did not run the risk of being persecuted (or suffering serious harm). However, Art. 8 of Council Directive 2004/83/EC provides that asylum (or subsidiary protection) may not be granted in such cases. This possibility remains at the discretion of Member States, and Italy decided not to avail itself of such a possibility, which was accordingly not transposed into Legislative Decree No. 251/2007 (Court of Cassation 9 April 2014, No. 8399, reasons for judgment).'

(2) Original version

Il secondo motivo denuncia violazione e/o falsa applicazione degli articoli 3 e 5 del decreto legislativo numero 251 del 2007, **censurando** la sentenza impugnata per avere omesso di applicare il principio dell'onere probatorio attenuato vigente nella materia, **non utilizzando** i propri poteri istruttori officiosi.

Pre-edited version

Il secondo motivo di ricorso riguarda la violazione e/o falsa applicazione degli articoli 3 e 5 del decreto legislativo numero 251 del 2007. La sentenza impugnata avrebbe omesso di applicare il principio dell'onere probatorio attenuato vigente nella materia. La Corte d'appello non avrebbe utilizzato i propri poteri di assumere prove d'ufficio.[3]

As seen in the examples above, dividing long sentences into shorter ones makes the inclusion of a subject necessary for each new sentence. This may lead, as in the case illustrated in Example 3 (from Order No. 15318/2020) below, to the repetition of the same subject (*il ricorrente*) or its substitution with a pronoun (*esso*).

(3) Original version

[. . .] **il ricorrente** aveva dichiarato di essere fuggito dal proprio Paese perché aveva ferito due trafficanti di alcool, armi e droga e, dopo aver denunciato il fatto al capo villaggio, che era a capo dell'organizzazione criminale, aveva subito violenza da parte di quest'ultimo; era stato anche denunciato alla Polizia per il ferimento dei trafficanti e del capo villaggio e temeva di rientrare nel proprio paese per timore di subire violenza e minacce da parte della Polizia;

Pre-edited version

Il ricorrente aveva dichiarato di essere fuggito dal proprio Paese perché aveva ferito due trafficanti di alcool, armi e droga. Dopo aver denunciato il fatto al capo villaggio, che era a capo dell'organizzazione criminale, **il ricorrente** aveva subito violenza da parte di quest'ultimo. **Il ricorrente** era stato anche denunciato alla Polizia per il ferimento dei trafficanti e del capo villaggio. **Egli** temeva di rientrare nel proprio paese per timore di subire violenza e minacce da parte della Polizia.[4]

---

3 'By the second ground of appeal, the appellant claims a breach of Articles 3 and 5 of Legislative Decree No. 251/2007. The Court allegedly did not apply the principle of burden of proof in force in such cases, which requires that the Court must use its powers for gathering evidence.'

4 'The appellant stated that he fled his country of origin because he had harmed there two traffickers in alcohol, arms and drugs. After reporting the matter to the village head, who was in fact at the head of the criminal organisation, the appellant had been assaulted by him. The appellant had also been reported to the police for the harming of the traffickers and the village head. He

5.1.2 Verb-subject → subject-verb

Another distinguishing feature of Italian judicial decisions is the verb-subject sequence, which results from the inversion of the much more common and unmarked subject-verb sequence. Whenever possible, in the pre-translation phase
the unmarked subject-verb sequence was used. In short sentences, such as in the
OV in Example 4 (from Judgment No. 538/2019), the intervention was minimal.

(4)	Original version	Pre-edited version
	È inammissibile il secondo motivo.	Il secondo motivo è inammissibile.[5]

In other cases, on the contrary, the introduction of the subject-verb sequence was
combined with other syntactic changes and forms of explicitation. For instance,
in Example 5 (from Judgment No. 2558/2020), in the PEV the subject-verb sequence
was kept in a separate sentence, while the subordinate clause introduced by
perché was transformed into an independent clause with the causal relation expressed using *infatti* and *della persona coinvolta* added for clarity purposes.

(5)	Original version	Pre-edited version
	[...] neanche può essere concessa la protezione umanitaria perché la situazione del Paese di provenienza esclude la sussistenza di una condizione di elevata vulnerabilità all'esito del rimpatrio, [...]	Neanche la protezione umanitaria può essere concessa. La situazione del Paese di provenienza non fa infatti emergere una condizione di elevata vulnerabilità della persona coinvolta in caso di rimpatrio.[6]

5.1.3 Impersonal or passive structures → personal and active structures

Italian judicial decisions also commonly feature impersonal or passive structures
with no explicit agent. While for an Italian lawyer the identification of implicit

---

feared returning to his country of origin and being subjected to the threats and violence of the
police.'
5 'The first ground is inadmissible.'
6 'Humanitarian protection cannot be granted either, as the situation in his country of origin is
not such as to put the appellant in a position of high vulnerability once he returns there. The
circumstances of vulnerability stated by the appellant cannot by themselves justify the granting
of a residence permit on humanitarian grounds.'

agents is straightforward, the same may not be true for either Italian non-lawyers or non-Italian lawyers who may not be familiar with Italian procedures. Therefore, impersonal or passive structures were frequently transformed into personal and active structures, which often required the explicitation of the agent and other modifications in the PEVs. In Example 6 (from Order No. 15318/2020), for instance, the impersonal *si contesta* became *il ricorrente contesta* and the OV one-sentence passage was split into four independent sentences, further entailing the explicitation of the agents as subjects (*Il ricorrente, Il Tribunale e la Corte d'appello di Catanzaro*). This, in turn, involved another frequent change made in the pre-translation phase, namely the transformation of embedded nominal constructs, whose lexical density is particularly high, into more diluted verbal constructs (*violazione del dovere di cooperazione istruttoria → hanno violato il dovere di cooperazione istruttoria; necessaria ai fini della valutazione della credibilità → avrebbero dovuto pertanto ascoltarlo in udienza al fine di valutare la credibilità delle sue dichiarazioni*) (see also below).

(6) Original version

[...] con il primo motivo di ricorso, **si contesta** l'omessa audizione del ricorrente in sede giudiziale in considerazione della carente verbalizzazione delle sue dichiarazioni innanzi alla Commissione Territoriale, con violazione del dovere di cooperazione istruttoria, necessaria ai fini della valutazione della credibilità; [...]

Pre-edited version

Con il primo motivo di ricorso, **il ricorrente contesta** la sua mancata audizione da parte dei giudici. Il ricorrente sostiene che la verbalizzazione delle sue dichiarazioni innanzi alla Commissione Territoriale era carente. Il Tribunale e la Corte d'appello di Catanzaro avrebbero dovuto pertanto ascoltarlo in udienza al fine di valutare la credibilità delle sue dichiarazioni. Non avendolo fatto, il Tribunale e la Corte d'appello hanno violato il dovere di cooperazione istruttoria.[7]

The OV in Example 7 (from Judgment No. 538/2019) consists in a single sentence of 247 words, which combines both passive and impersonal structures. After splitting the

---

7 'By the first ground of appeal, the appellant argues that he was not granted a hearing by the judges and also maintains that the report of his statements before the Territorial Commission was poor. Thus, the Catanzaro Court of first instance and Court of Appeal should have summoned him for a personal interview to properly assess the credibility of his statements. By not doing so, both the Court of first instance and the Court of Appeal breached the duty of inquiry.'

original sentence into six sentences, other major interventions were necessary to reduce the obscurity and increase the readability of the passage. The passive structure in the past participle clause (*non tipizzati o predeterminati, neppure in via esemplificativa, dal legislatore*) was transformed into a more explicit, active relative clause, which also involved the recasting of the agent (*che il legislatore non elenca, neppure in via esemplificativa*). Yet, more interesting is the next agentless passive structure in the OV, which is immediately followed by a gerund with an implicit agent (*non può cioè essere in nessun caso elusa la verifica della sussistenza di una condizione personale di vulnerabilità, occorrendo dunque una valutazione individuale, caso per caso . . .*). In this case, both the passive and the gerund were converted into active forms, with the explicitation of the agent as their subject, which ultimately led to the creation of two separate sentences (*Il giudice non può in nessun caso eludere la verifica della sussistenza di una condizione personale di vulnerabilità. Egli deve svolgere una valutazione individuale, caso per caso ...*).

(7)   Original version

La pronuncia è così conforme al principio ribadito – senza che sorga questione in ordine alla verifica dell'applicabilità del decreto-legge 4 ottobre 2018, n. 113, convertito con modificazioni dalla legge 10 dicembre 2018, n. 132 – secondo cui il diritto alla protezione umanitaria è in ogni caso collegato alla sussistenza di «seri motivi», **non tipizzati o predeterminati, neppure in via esemplificativa, dal legislatore,** cosicché essi costituiscono un catalogo aperto, tutti accomunati dal fine di tutelare situazioni di vulnerabilità individuale attuali o pronosticate in dipendenza del rimpatrio: **non può cioè essere in nessun caso elusa la ver-**

Pre-edited version

La pronuncia è così conforme al principio [OMISSIS] secondo cui il diritto alla protezione umanitaria è collegato alla sussistenza di «seri motivi», **che il legislatore non elenca, neppure in via esemplificativa.** Si tratta di un catalogo aperto di motivi, tutti accomunati dal fine di tutelare situazioni di vulnerabilità individuale attuali o prevedibili in caso di rimpatrio. **Il giudice non può in nessun caso eludere la verifica della sussistenza di una condizione personale di vulnerabilità. Egli deve svolgere una valutazione individuale, caso per caso, della vita privata e familiare del richiedente, comparata alla situazione personale che egli ha vissuto**

ifica della sussistenza di una condizione personale di vulnerabilità,
occorrendo dunque una valutazione individuale, caso per caso,
della vita privata e familiare del richiedente, comparata alla situazione personale che egli ha vissuto
prima della partenza e alla quale
si troverebbe esposto in conse-
guenza del rimpatrio: [...]

prima della partenza e alla quale
si troverebbe esposto in caso di
rimpatrio. [...][8]

5.1.4 Nominal style → verbal style

As already mentioned with regard to Example 6, Italian judicial decisions are characterised by a high degree of lexical density (Ondelli 2013: 77) and a low level of readability. In Example 8 (from Order No. 15318/2020), a relative clause was transformed
into a separate sentence and the head of the relevant complex noun phrase was
transformed into a verb (*potere officioso del giudice* → *spetta al giudice*).

(8)   Original version
      [...] detta interpretazione era, del resto
      consolidata nella giurisprudenza di
      questa Corte, che ha affermato come
      non sia ravvisabile una violazione
      processuale sanzionabile a pena di
      nullità nell'omessa audizione per-
      sonale del richiedente, atteso che il

      Pre-edited version
      Questa interpretazione era, del resto,
      consolidata nella giurisprudenza di
      questa Corte, la quale ha affermato
      che la mancata audizione del richie-
      dente non costituisce una violazione
      delle regole procedurali sanzionata
      con la nullità degli atti. Infatti, questa

---

8 'Thus, the decision complies with the principle [...] according to which the right to humanitarian protection must always be based on «serious reasons» which cannot be predefined, not even
by way of examples, by the legislator, but constitute instead an open catalogue, where they all
share the common goal of addressing conditions of individual vulnerability, which are either already existing or may possibly result from the applicant's return to his/her country of origin.
This means that an assessment of the existence of a personal condition of vulnerability must always be carried out by the Court. Such an individual case-to-case assessment of the applicant's
private and family circumstances should then be compared to his/her living conditions in the
country of origin before departure and to those he/she would be exposed after returning there.'

rinvio, contenuto nell'art. 35, comma 13, del d.lgs. n. 25 del 2008, al precedente comma 10 che prevede l'obbligo di sentire le parti, non si configura come un incombente automatico e doveroso, ma come un diritto della parte di richiedere l'interrogatorio personale, **cui si collega il potere officioso del giudice d'appello di valutarne la specifica rilevanza** (Cassazione civile sez. VI, 07/02/2018, n. 3003; in senso conforme Cass. Civ., sez. 06, del 21/11/2011, n. 24544); [...]

Corte aveva già stabilito che l'audizione delle parti non è un obbligo del giudice, ma un diritto della parte, la quale può chiedere l'interrogatorio personale. **Spetta poi al giudice d'appello valutare la rilevanza dell'audizione** (Cassazione civile sez. VI, 07/02/2018, n. 3003; in senso conforme Cass. Civ., sez. 06, del 21/11/2011, n. 24544);[9]

In Example 9 (from Judgment No. 23757 2019), on the contrary, the head noun remained unchanged (*esame*), but the explicitation of the phrase starting with *in ragione di* required a series of other major modifications. Firstly, the whole phrase was brought forward, with the agent (*La Corte di appello*) as its subject and the head noun *esame* as the object. This, in turn, implied the introduction of a verb (*ha svolto*) and the normalisation of another typical syntactic feature of Italian judicial drafting, namely the adjective-noun inversion (*puntuale ed autonomo esame* → *esame attento*), with simplification of the adjective compared to the OV.

(9) Original version

La Corte di merito ha ritenuto l'insussistenza di condizioni di vulnerabilità legate a fattori soggettivi o desumibili dalle condizioni politico-sociali del paese di origine **in ragione di un puntuale ed autonomo esame**,

Pre-edited version

La Corte di appello ha svolto un esame attento. Essa ha escluso che sussistano condizioni di vulnerabilità, legate alla situazione soggettiva del ricorrente o alle condizioni politico-sociali del paese di origine. Per

---

9 'Such an interpretation was already well-established in the case law of this Court, which held that the failure to grant the applicant a hearing does not constitute a breach of the procedural rules to be sanctioned with the nullity of the proceedings. Indeed, this Court had already established that the hearing of the parties is not a duty of the court, but a right of the party, who can ask for a personal interview. It is then up to the court hearing the appeal to decide on the relevance of such a hearing (Court of Cassation, Civil Section VI, 7 February 2018, No. 3003; Court of Cassation, Civil Section VI, 21 November 2011, No. 24544).'

escludendo che l'azionata pretesa fosse riconducibile al catalogo "aperto" della protezione umanitaria.

tale ragione, ha ritenuto che la situazione del ricorrente non dia diritto alla protezione umanitaria.[10]

5.2 Terminology and phraseology

As seen so far, the interventions on the syntactic structure of the original texts were meant to reduce the complexity and increase the readability of the source texts, and ultimately facilitate the translation process. These interventions often led to a higher explicitness in the pre-edited versions. However, simplification and explicitation were also adopted in relation to the terminology and the phraseology used in the OVs. Indeed, Italian judicial decisions are characterised by terms and phrasemes which not only are often described as opaque, cryptic, elusive (Cordero 1988: 310), but are also archaisms preserved almost exclusively in judicial, legislative or administrative drafting.

In Example 1 above, for instance, two possibly obscure phraseological units were normalised (*in virtù di* → *quando, atteso che* → Ø), but also an archaic adverb was substituted with a much more common subordinating phrase (*ove* → *nella quale*). Likewise, in Examples 10 (from Order No. 11170/2020) and 11 (from Judgment No. 538/2019) below, the conjunctions *onde* and *sicché*, typical of the Italian language of law and public administration, were substituted with the more genre-independent conjunctions *dunque* and *quindi*, respectively.

(10) Original version

Il diritto alla salute del richiedente integra, dunque, un diritto umano fondamentale, **onde** appare necessario accertare se, considerata la natura e gravità della malattia del richiedente, la necessaria terapia

Pre-edited version

Il diritto alla salute del richiedente costituisce un diritto umano fondamentale. È **dunque** necessario accertare se, considerata la natura e gravità della malattia del richiedente, la necessaria terapia medica o farmacologica

---

10 'The Court of Appeal carried out a proper assessment and ruled out the existence of any conditions of vulnerability linked to the appellant's subjective situation and to the political and economic situation of his country of origin. For this reason, the Court of Appeal held that the appellant's situation did not entitle him to humanitarian protection.'

medica o farmacologica possa essergli somministrata anche nel suo
paese di origine.

possa essergli somministrata anche nel
suo paese di origine.[11]

(11) Original version
Sicché il ricorrente ha frainteso la
ratio decidendi.

Pre-edited version
Il ricorrente ha quindi frainteso la **motivazione della sentenza impugnata**.[12]

Example 11 is also interesting because of a frequent terminological feature of Italian judicial decisions, namely the use of Latinisms, which is also illustrated in Example 12 (from Judgment No. 23757/2019). Although some Latinisms are used both
in Italian and English (e.g. as in the case here of *ratio decidendi*), in the pre-
translation phase the lawyers decided to replace any Latinisms which may lead
to involuntary associations with foreign case law (the *ratio decidendi* is an essential element of judicial precedents in Common Law) with their Italian equivalents
with a higher informative potential which may be found in relevant European
acts. For example, *valutazione* and *individuale* appear in close connection in Directive 2011/95/EU on standards for the qualification of third-country nationals or
stateless persons as beneficiaries of international protection.

(12) Original version
Nell'impugnata sentenza sarebbe
stato violato l'art. 32, comma 3-bis,
d.lgs. n. 25/2008 che, espressamente,
impone una valutazione **ad per-
sonam** in ordine alla sussistenza
di ragioni umanitarie.

Pre-edited version
Il ricorrente sostiene altresì che la
sentenza impugnata viola l'art. 32,
comma 3-bis, d.lgs. n. 25/2008 che,
espressamente, impone una valutazione **individuale** in ordine alla
sussistenza delle condizioni per il
rilascio di un permesso per ragioni
umanitarie.[13]

One final intervention worth mentioning relates to the terminology found in the judicial decisions analysed, which mainly relates to three broad fields: the requisites

---

11 'The applicant's right to health is a fundamental human right. It is therefore necessary to assess if, considering the nature and seriousness of the applicant's illness, the necessary medical or
drug treatment can be administered also in his country of origin.'
12 'Thus, the appellant misunderstood the reasons for the appealed judgment.'
13 'Furthermore, the appellant claims that the challenged decision breaches Art. 32(3-*bis*) of Legislative Decree No. 25/2008, which explicitly requires an individual assessment of the requirements for the granting of a humanitarian residence permit.'

third-country nationals or stateless persons must meet in order to obtain international protection; the administrative procedure to obtain it; and the judicial procedure to challenge a decision rejecting such protection. Although some terms (and procedures) may not be of immediate understanding to non-lawyers, such as the linguists and translators involved in the project, in most cases the terminology used – usually consistently – in the judicial decisions under examination did not undergo any particular modification in the pre-edited version (the main exception being the synonyms used in the OVs to refer to courts which have been replaced by one single term in PEVs; see Challenges and reference documentation above). Instead, the linguists and translators involved in the TrIACLE project remitted their doubts about the terminology used to the lawyers and linguists, who solved them either in writing using comments in the source texts or in online meetings (see Methodology above).

6 Conclusions

Italian judicial decisions are characterised by linguistic features that are generally considered to be deviations from the norm and may seem unusual even to non-lawyers who are Italian native-speakers. These features contribute to the complexity and bombastic effect of such decisions and ultimately to their lack of readability, which makes translating them a particularly challenging task. Considering that the texts translated by the TrIACLE team are meant for non-Italian-speaking international legal practitioners of EU Member States other than Italy, at the outset of the project the team decided that the translations needed to be as informative as possible so as to minimise the readers' processing effort. However, instead of producing clearer, plainer texts directly from the original versions of the selected decisions, the team opted for a pre-editing phase in which an intra-lingual translation was carried out. By applying simplification and explicitation strategies at the morphosyntactic and lexical (mainly phraseological but also terminological) levels, complex, convoluted, or even obscure passages were transformed into more accessible and legible, as well as syntactically less norm-deviating, passages in Italian.

The examples provided in this paper illustrate the impact on the original texts of the intermediate phase of intra-lingual translation, which to our knowledge has gone largely unnoticed in translation studies in general, and legal translation studies in particular. Indeed, while some – admittedly still scarce – research has been conducted on the use of plain language principles in the direct production of target texts (Adams 2005; Toledo Báez and Conrad 2017; Williams 2013), very little has been written on the application of plain language – and thus of simplification and explicitation strategies – in intra-lingual translation preparatory to the actual production phase (Sciumbata 2018). The aim of this paper was therefore to bring to the fore how intra-lingual translation can support the understanding of the original text and ideally help the production of more accessible target texts. However, the observations made in this paper are limited due to two main reasons. First of all, they are restricted to a very small number of texts all belonging to the same text type. This means that further research is needed, first, to investigate whether intra-lingual translation is a phase in other inter-lingual translation projects and, second, to verify whether similar simplification and explicitation strategies are applied in relation to other text types. The second limitation is that, due to space constraints, the observations do not account for the impact of intra-lingual translation on the translated texts. Again, further research is called for to assess such impact and to explore the possible modifications required by the transfer from an intermediate source-language text to a final target-language text.

References

Adams, Matthew. 2005. Plain English: The cure against translating infectious legal-speak. *ATA Chronicle* 34(9). 28–29.

Associazione Italiana Traduttori e Interpreti. 2013. "Codice di deontologia e di condotta". http://www.aiti.org/associazione/codice-deontologico (accessed 21 February 2022)

Bhatia, Vijay K., Jan Engberg, Maurizio Gotti & Dorothee Heller (eds.). 2005. *Vagueness in Normative Texts*. Bern/Frankfurt am Main: Peter Lang.

Cao, Deborah. 2007. *Translating Law*. Clevedon/Buffalo/Toronto: Multilingual Matters.

Cordero, Franco 1988. *Stilus curiae (Analisi della sentenza penale)*. *Atti Del Convegno Internazionale per l'inaugurazione Della Nuova Sede Della Facoltà, Ferrara 10–12 Ottobre 1985*, 293–312. Padova: CEDAM.

Cortelazzo, Michele. 1995. Lingua e diritto in Italia. Il punto di vista dei linguisti. In Leo Schena (ed.), *La lingua del diritto. Difficoltà traduttive. Applicazioni didattiche. Atti del primo convegno internazionale, 5–6 ottobre 1995*, 35–50. Milano: CISU.

Cronin, Michael. 2003. *Translation and Globalization*. London/New York: Routledge.

Engberg, Jan &Dorothee Heller. 2008. Vagueness and indeterminacy in law. In Vijay K. Bhatia, Christopher N. Candlin & Jan Engberg (eds.), *Legal Discourse across Cultures and Systems*, 145–168. Hong Kong: Hong Kong University Press.

European Parliament. 1999. "Tampere European Council 15 and 16 October 1999. Presidency Conclusions". https://www.europarl.europa.eu/summits/tam_en.htm#a (accessed 19 February 2022)

Gialuz, Mitja, Luca Lupària & Federica Scarpa (eds.). 2014. *The Italian Code of Criminal Procedure. Critical Essays and English Translation*. Padova: Wolters Kluwer Italia/CEDAM.

Gialuz, Mitja, Luca Lupària & Federica Scarpa (eds.). 2017. *The Italian Code of Criminal Procedure. Critical Essays and English Translation. With a Preface by the Italian Minister for Justice.* 2nd edn. Padova: Wolters Kluwer Italia/CEDAM.

Gile, Daniel. 1986. La traduction médicale doit-elle être réservée aux seuls traducteurs-médecins? Quelques réflexions. *Meta* 21(1). 26–30.

ISO 1700. 2017. Translation services – Requirements for translation services, International Organization for Standardization.

Joseph, John E. 1995. Indeterminacy, translation and the law. In Morris Marshall (ed.), *Translation and the Law*, 13–36. Amsterdam/Philadelphia: John Benjamins.

Lafeber, Anne. 2018. The skills required to achieve quality in institutional translation: the views of EU and UN translators and revisers. In Fernando Prieto Ramos (ed.), *Institutional Translation for International Governance: Enhancing Quality in Multilingual Legal Communication*, 63–80. London/ New York: Bloomsbury Academic.

Ondelli, Stefano. 2013. Un genere testuale oltre i confini nazionali: la sentenza. In Stefano Ondelli (ed.), *Realizzazioni testuali ibride in contesto europeo. Lingue dell'UE e lingue nazionali a confronto*, 67–91. Trieste: EUT Edizioni Università degli Studi di Trieste.

Ondelli, Stefano. 2014. Drafting court judgements in Italy: history, complexity and simplification. In Vijay. K. Bhatia, Giuliana Garzone, Rita Salvi, Girolamo Tessuto & Christopher Williams (eds.), *Language and Law in Professional Discourse: Issues and Perspectives*, 29–45. Cambridge: Cambridge Scholars Publishing.

Pokorn, Nike. 2005. Challenging the Traditional Axioms. Amsterdam/Philadelphia: John Benjamins.

Pokorn, Nike. 2009. Natives or Non-natives? That is the Question . . . Teachers of Translation into Language B. *The Interpreter and Translator Trainer* 3(2). 198–208.

Pym, John. 2015. Managing Translations and Assuring Quality at CERN. Round-table presentation delivered at the Transius Conference, University of Geneva, 26 June, 2015.

Rega, Lorenza. 1997. La sentenza italiana e tedesca nell'ottica della traduzione. *La Lingua Del Diritto: Difficoltà Traduttive, Applicazioni Didattiche: Atti Del Primo Convegno Internazionale, Milano, 5-6 ottobre 1995*, 117–126.

Šarčević, Susan. 1997. *New Approach to Legal Translation*, The Hague/London/Boston: Kluwer Law International.

Scarpa, Federica. 2020. *Research and Professional Practice in Specialised Translation*. London: Palgrave-Macmillan.

Scarpa, Federica & Alison Riley. 1999. *La traduzione della sentenza di common law in italiano*. Trieste: EUT Edizioni Università di Trieste.

Scarpa, Federica,Katia Peruzzo & Gianluca Pontrandolfo. 2017. Methodological, terminological and phraseological challenges in the translation into English of the Italian Code of Criminal Procedure: What's new in the Second Edition. In Mitja Gialuz, Luca Lupària & Federica Scarpa (eds.), *The Italian Code of Criminal Procedure. Critical Essays and English Translation*. 2nd edn, 53–80. Padova: Wolters Kluwer Italia/CEDAM.

Sciumbata, Floriana. 2018. Un'esperienza di insegnamento tra plain language e traduzione. *Rivista Internazionale Di Tecnica Della Traduzione/International Journal of Translation* 20. 195–207.

Toledo Báez, Cristina & Claire Alexandra Conrad. 2017. Informational pamphlets for asylum seekers in English. *Revista Española de Lingüística Aplicada/Spanish Journal of Applied Linguistics* 30(2). 559–591.

Williams, Malcolm. 2013. Plain Language Translation: Principles and Techniques. *FORUM* 11(2). 201–230.

Farida Buniatova

Screening the law: Popularization practices in "How To Get Away With Murder"

1 Introduction

Nowadays, mass media has become an effective tool to disseminate information and popularize knowledge pertaining to many important spheres of human life. Not only does mass media inform us about what we supposedly do not know about the world around us, but also shapes our perceptions and forms images in our heads.

Law is one of the areas increasingly popularized by mass media via TV programs, the so-called courtroom shows and legal dramas, movies, and TV series centered around law, crime, and the courtroom. As for TV shows, they can be seen as a "window on the world," and a way to provide people with information they would not otherwise be able to obtain (Brown Graves 1999).

Research (Sherwin 2006) shows that a rapid sequence of images, as when we watch movies and TV series, prevents critical thinking and therefore, contributes to perceiving and retaining the visual message without analyzing it. According to statistics, in 2021 an average American spent about three hours and 17 minutes watching TV each day[1] (Statista 2022). More importantly, however, this figure has been falling in recent years and, according to forecasts (Statista 2022), will continue to fall in the years to come.[2] However, those figures do not imply that people are spending less time in front of the television screen, as they now tend to spend more and more of their time using their cell phones, tablets and laptops while opting for streaming services such as Netflix, HBO, Hulu, Amazon Prime and many others. Thus, according to the statistics, as of March 2021, most of the respondents in the US (Statista 2022) watched from three to five hours of content through a video streaming service such as Netflix or HBO Max a day, while 14% of the respondents stated that they view content on Video on Demand platforms for 12 hours or more (Statista 2022).

---

1 The exact figure depends on a region, e.g., an average person in Asia spends 2 hours and 56 minutes, in Europe – 3 hours and 49 minutes. Average daily on-demand TV and video viewing time in selected countries worldwide as of October 2018, by age group. Retrieved April 29 2022, from Statista.
2 Though there was an exception in 2020 due to the COVID-19 pandemic, when media consumption saw a sharp increase.

https://doi.org/10.1515/9783111048789-009

When people spend so much time watching movies, TV shows, and various other broadcasts, they are obviously confronted with stereotypical images of minorities, ethnic groups, representatives of different jobs and professions, etc. – something that is more likely done by movie directors and producers subconsciously rather than intentionally (Heinrich Böll Stiftung 2010). Indeed, research has revealed that the media affects a person's perception of their social groups and inter-group relations (Gerbner et al. 1980: 46). For instance, evidence has shown that stereotypical portrayals of Afro-Americans negatively affect their self-esteem (Tan and Tan 1979). Despite a modern trend towards a more realistic portrayal of certain minorities, professions, persons, and cultures; the predominance of stereotypes and bias persists (Brown Graves 1999).

Mass media also shapes human expectations about the justice system, legal procedure, and the image of a lawyer. This leads to an important issue of whether mass media needs to be accurate and aimed at teaching law, or rather, to serve the purpose of entertaining the viewers and impart a dream world of justice, fairness, and order.

Over the past few decades, there have been many legal dramas and series on television and on streaming services. It is worth remembering *L.A. Law* (1986-1994), which romanticized the American legal system, centering around a Los Angeles law firm, which dealt with 'hot-button' issues such as, inter alia, racism, domestic violence, and abortion. One of the most well-known legal dramas is *Law and Order* (1990 to present), which depicts the investigations of deadly crimes, gathering evidence, questioning suspects, and proving guilt beyond a reasonable doubt in a courtroom. Another popular series is *Suits* (2011-2019), which, by contrast, concentrates on the work of a legal firm, rather than on investigating real cases and depicting courtroom proceedings.

This paper is aimed at studying the image of a Black criminal defense attorney and a university professor, Annalise Keating, represented in the popular American TV drama *How To Get Away With Murder* aired on ABC from September 2014 to May 2020 and comprising six seasons and 90 episodes in total. The TV series, which can as well be qualified as a legal thriller, differs from most other legal dramas in terms of the variety of social justice themes, diverse cast, and the extraordinary lead performance of Viola Davis. The show touches upon such social issues as alcoholism, interracial relationships, gay marriage, HIV, the mass incarceration of Black men, and law enforcement corruption. Moreover, it depicts the way the characters' morals are tested and questions if anyone, under the relevant circumstances, could be capable of committing a violent crime.

The issues investigated in the present paper are whether the image depicted in the series represents a realistic picture of what the life of law students is like in the US and whether the image of Annalise Keating is subject to racial stereotypes and corresponds to the real image of a university professor.

The study provides a thorough analysis of the law students' life at a prominent American university and of the image of Annalise Keating, the protagonist, who is both a prominent attorney dealing with cases of utmost complexity, and a professor at a prestigious Philadelphia law school.

2 Popularization of law students' life

> One of the complications of teaching criminal law is
> [the students] have watched a lot of the shows
> and think they know what the laws are.
> (Paul Robinson
> Criminal Law Professor
> at University of Pennsylvania)

The central character in the TV series, *How To Get Away With Murder,* is Annalise Keating, who is not simply a learned scholar but an outstanding criminal defense attorney helping her students to achieve advanced trial skills. Students struggle to get a seat in her class and even transfer from other law schools to be able to study under her guidance.

The group of students attending Keating's class is rather diverse in terms of race, and this highlights the idea of the US as a post-racial society. However, let us point out that student populations of most US universities, and particularly selective ones, are still disproportionately White (Supiano 2015). Even though US universities, and especially law schools, are becoming more diverse, 70%-90% of law students are still classified as White (Torres-Spelliscy, Chase, and Greenman 2010).

The series was shot in the University of Southern California, which, like most Californian universities, has one of the most diverse student populations in the whole country. However, even according to formal data provided by the Data USA website (DataUSA 2022), the enrolled student population at the University of Southern California is 29.5% White, 18.7% Asian, 15% Hispanic or Latino, 5.48% Black or African American, 3.89% Two or More Races, 0.264% Native Hawaiian or Other Pacific Islanders, and 0.171% American Indian or Alaska Native. The most drastic contrast can be observed among doctoral students, who are represented by 49.8% White, 15.1% Hispanic or Latino, and 9.74% Black or African American students (DataUSA 2022).

Thus, racial diversity in Professor Keating's classroom does not correspond to the reality of American law schools, especially ones at elite universities, one of which is depicted in the series. Therefore, the issue of racial disparities of the American higher education system in *How To Get Away With Murder* is highly disregarded.

It might be argued (Sholat and Stam 2014), however, that the over-representation of racial minorities in the series is intentional and aims at showing an ideal post-racial society where all minorities and races are equally represented. This proves to be wrong according to the authors of the series, who claim that the series mirrors the real world, and that the United States is a post-racial nation (Enlow 2015):

> *In ShondaLand our shows look like the world does . . . To me, that was not some difficult, brave, special decision I made. It was a human one, because I am a human . . . This is not the Jim Crow's south, we're not ignorant, so why wouldn't we do that? I still can't believe I get asked about it [the racial diversity in her casts] all the time, as if being normal, TV looking like the normal world, is an innovation.*

In one of his interviews, Peter Nowalk, creator and executive producer of *How To Get Away With Murder*, admitted that the issue of race is one of the most important ones in the series: *"Themes I want to play with . . . are race and class, especially in an elite university. We're not colorblind to who these people are; their identities . . . and everything about them will become a potential story"* (Prudom 2014).

As for the character of Professor Keating, she inspires awe and students obviously fear her while she, from the very beginning, demonstrates her harsh and tough approach to teaching.

> *This is a sacrifice. From this point on, you will have no time for friends or family. Instead, you wake up hating yourself for choosing this life, but you'll get up anyway, killing yourself to win cases . . . only to lose and watch innocent people go to jail. And then you'll drink to make yourself feel better, or take pills, or fantasize about going to sleep . . . forever. That's the life you're choosing. Brutal, mean, depressing, ruthless . . . but that's what it costs to change the world. So who wants in?*

At the very first lecture, she asks students what their "legal passion" is and, if a student fails to give a quick and sensible response, they are ejected from the course. As for a real-life situation, one can hardly imagine that any institution, let alone a top-ranking Philadelphia law school, can allow its faculty to dismiss a student from a course for failing to answer a question on the first day of studies. The mere fact of enrollment does not bear much significance in the series as students must earn a right to be part of Professor Keating's course. Moreover, she decides, according to her own criteria, whether a student deserves a seat or not, and in the latter case, makes unwelcome ones withdraw from the course. It seems rather doubtful that any professor would be authorized to force a student to drop out.

However, Professor Keating's harsh approach towards her students makes the show more entertaining and intense.

Another aspect that grasps attention is the time that students allocate to actual studies. Students in the series seem to have plenty of free time, which is obviously not the reality of studying for the legal profession. For instance, in a Philadelphia Inquirer article (Philadelphia Inquirer 2014), Penn students called the show "over-the-top and sometimes unrealistic". The review of the UPenn Biddle library simply inquires *"[. . .] we do have one comment regarding whether or not the series is realistic: Why is no one ever studying in the library?!"* (Penn Law 2014).

Indeed, the group of five students, which Professor Keating has selected as the most outstanding ones, seems to do nothing except struggle to win the "trophy" she promised to the 'smartest' student. This contradicts the apparent reality where university classes in Criminal Law, Torts, Contracts, and the like, are compulsory and therefore, students are not allowed to skip them in lieu of work.

Furthermore, in a law school, Criminal Law, Criminal Procedure and Trial Advocacy are different classes; however, in the series, all three subjects are taught within Professor Keating's Criminal Law class. This is rather contradictory as Criminal Procedure and Trial Advocacy are generally second- and third-year classes.

Moreover, Professor Keating focuses her Criminal Law class solely on the topic of murder as if it were the essence of the subject. Thus, the series creates an impression that mastering Criminal law is limited to studying the issue of murder.

Apart from that, Professor Keating begins her first class by saying, *"Unlike many of my colleagues, I will not be teaching you how to analyze the law or theorize about it"*, thus contradicting the essential principles of legal education, especially with regard to first-year students, who simply cannot skip hundreds of years of American jurisprudence and completely ignore the curriculum.

In another episode, Professor Keating makes an arrangement with students to meet her in the courthouse at 9:00 am. When one of the students objects, and points out that they have morning classes that day, Professor Keating responds that she could not care less – *"The way you're whining right now makes me believe you think I'm your mother, Ms. Pratt. Show up tomorrow or drop out of the competition. It's that simple"*.

The students in the series are engaged in the work of a legal clinic which, indeed, is an effective way to develop legal skills. However, the series shows an unrealistic picture of students gathering evidence and consulting for high-profile cases, thus creating a false impression about the life of a law student. In fact, law schools do not offer training; they provide legal education and teach legal skills that can be further applied in actual practice. Neither are students normally allowed to study actual open cases, while the only thing Professor Keating's students do is work on her current cases which seems to be unheard of in a real-life class.

Another unlikely scene is where Professor Keating invites her entire class to her law firm to speak to defendants who reveal their stories to a group of unaffiliated and unemployed students, thus ignoring and waiving their attorney-client privilege.

The aforementioned case is not the only one where Professor Keating's students completely ignore ethical standards. Throughout the show, they falsify documents, pretend to be attorneys to get access to prisoners, and so on. However, in reality, law students spend so much money and time on school and studies, that they are paranoid about making even the smallest of mistakes that could result in their failure to become a lawyer. In fact, second-year students are supposed to take a course in Professional Responsibility or Ethics. Moreover, to be admitted to practice, apart from the Bar examination, students must pass the Multistate Professional Responsibility Exam to prove they are aware of, and adhere to, ethical standards, i.e. matters such as client care, conflict of interest, confidentiality, dealing with client money, etc. Therefore, not only is it unethical, but in some US states, falsely claiming to be a lawyer amounts to a felony. For instance, the Texas Penal Code provides that a conviction for falsely claiming to be a lawyer is punishable as a third-degree felony, with a maximum fine, under Texas state law, of up to $10,000 and a custodial sentence of up to 10 years (38 Tex. Penal Code §122 (b) (2019)).

To sum up, the conclusion that might be drawn is that *How To Get Away With Murder* presents a rather distorted image of the reality of being a law student. Indeed, a show accurately depicting the routine of law students, consisting of reading heavy old books, spending hours in a library, and memorizing material, would be rather dull and annoying. However, it certainly creates a false image in people's minds, and could be capable of affecting teenagers' choice of profession which, having turned out to be quite the opposite to what they had seen in the show, might lead to disillusionment and frustration.

3 Image of annalise keating as a black woman attorney

> *I did not want a character that fit into the quote-end quote network TV parameters.*
> *I wanted to be more rooted in reality and in life,*
> *and more rooted in the complexity of what it means to live a life.*

Viola Davis
(Turchiano 2020)

In the mid-1960s, after the passing of the Civil Rights Act (1964) and the Voting Rights Act (1965) prohibiting discrimination, "the coming of television pushed the film industry, which was competing with the new medium in the era of the baby boom, towards greater revelation, controversy, and maturity, including in the realm of race" (Scott 2016: 10). In the years that followed American television started to employ various strategies in order to prove a lack of bias and discrimination, the most popular being "diversity casting, which showcases a multicultural cast without acknowledging or addressing cultural and social differences" (Nilsen and Turner 2014: 5).

One of the important milestones in the development of color-blind trends and post-racial politics was the presidency of Barack Obama (2009-2017), who "became the ultimate embodiment of the America's post-racial ideals" (Martens and Povoa 2017: 120). Since then, American television has started to release drama series featuring more and more minority representatives.

One of the most prominent examples of this is *How To Get Away With Murder*, which features a racially diverse cast. Apart from Professor Keating, who is Afro-American, there is her lover (Nate), who is a Black police officer, and her group of student trainees, which includes a diverse crowd of two Black American students, namely Haitian American (Wess) and Afro-American (Michaela), as well as a Latino (Laurel).

As for Viola Davis, who plays the lead role of Professor Keating, she became only the second Black actress to star in a prime-time American network series in almost forty years (Nilsen and Turner 2014) and the first African-American actress to win an Emmy for Outstanding Lead Actress in a Drama Series. In her interview for *The Hollywood Reporter* (Howard 2015), Viola Davis stated that she has a darker skin color than most other Black actresses in Hollywood and was almost fifty years old when she started working on the series, which is usually considered "too old" and "not sexy enough" for Hollywood:

> *I had no precedent for this role. I've never seen anyone, 49-year old, dark-skinned, woman, who is not a size 2, be [in] a sexualized role on TV, film, anywhere, ever. And then suddenly this role came to me. But to say it was fear would be an understatement; it was bigger than fear. . . . And then my big "a-ha" moment was: "this is your moment to not typecast yourself."*

Professor Keating is a prominent case-winning lawyer and the show aims at showing how a Black lawyer can survive in contemporary American society, while highlighting how caring and strong this person might be. Professor Keating's representation as a Black female attorney is extremely significant as it is one of the very few images of a Black woman lawyer on screen and it could therefore have an impact on viewers' perceptions about women of color in the legal profession.

From the very first episode, we observe Professor Keating as a busy and powerful female professional. In episode one she introduces herself to students and announces that the name of her class "Criminal Law 100" is referred to as 'How To Get Away With Murder'. Immediately afterward, she voices three key elements needed to cause doubt in a case and to therefore be able to get away with murder: 1. Discredit witnesses; 2. Introduce a new suspect; and 3. Bury the evidence.

The episode is shocking because it is highly unusual for a woman of color to be a professor who explicitly teaches her class how to win a case by methods that can hardly be considered ethical. Startling is the fact that Professor Keating instructs her students how to discredit people, shift blame, and hide evidence; thus, showing that an attorney must do their job well irrespective of the means, whether ethical or not.

However, the show underlines stereotypes associated with Black people in the US and upheld by a White majority. As certain authors (Pettigrew and Meertens 1995: 62; Essed 1990) claim, prejudices today are not expressed, but rather implied and indirect, or as they are referred to – "latent prejudice" or "everyday racism". This new kind of prejudice is a system of ethnic dominance along two dimensions (van Dijk 2012) – the social dimension and the cognitive one. The social dimension, according to van Dijk, includes everyday social practices of discrimination against minorities and ethnically different groups, including discourse. The cognitive dimension refers to stereotypes, prejudices, and ideologies of minorities and ethnic groups. Both lead to discrimination against minorities and ethnic groups. According to certain studies, for instance, Black people are characterized as "athletic" and "rhythmic" (which are positive characteristics) whilst being "low in intelligence" (Rodriguez 2019: 22) (obviously negative).

All these stereotypes and prejudices are significant for the media and television, who deliver them to an audience who may never have had direct contact with persons belonging to a certain minority or ethnic group, therefore shaping people's perceptions of minorities and race. Studies reveal that African Americans, or people with darker skin, depicted as suspects in lawsuits in TV shows, are more commonly perceived by viewers to be guilty than are any White suspects (Hurwitz and Peffley 1997).

Latent stereotypes are also reflected in depicting Professor Keating in *How To Get Away With Murder*. Firstly, she is shown as a typical *Black Mammy*, who is loyal and faithful to her *White* family, represented by Professor Keating's students.

The Mammy was a large, independent woman with pitch-black skin and shining white teeth (Jewell 1993), who raised her White misters' children and loved them unconditionally. She was wise and often gave harmless, while smart, advice to her mistress. However, she dominated and controlled her own husband and children being a quasi-tyrant in her own family setting (Jewell 1993).

Likewise, Professor Keating is an all-knowing, all-seeing, and all-understanding figure foreseeing her enemies' moves and caring for her clients and students. Even when her students kill her husband, she does not report it to the police, but devises a story to help her students avoid prosecution. She lies to the police and becomes an accomplice in her husband's murder in order to protect her *White* family of students.

Also, Professor Keating resembles the character of *Sapphire* as portrayed in the popular "Amos'n'Andy" sitcom about Black characters set in Chicago and Harlem, New York City. Sapphire was a bossy, headstrong woman engaged in a continuous verbal battle with her husband, Kingfish (Jewell 1993). Her fierce independent domination over her husband contributed to her image of a matriarch. In the "Kill me, Kill me, Kill me" episode of *How To Get Away With Murder,* Professor Keating confronts her husband immediately after learning about his extramarital affair. She angrily insults him and behaves aggressively, boasting about her own affair with Nate, and so demonstrating her hard, dynamic, and mysterious personality.

Furthermore, Professor Keating's image might be construed as the one of the modern *Jezebel.* According to the Bible, Jezebel was a queen who was married to King Ahab of Israel and persuaded him to introduce the worship of the Tyrian god Baal-Melkart, a nature god. She has come to be known as an archetype of the wicked woman (Britannica 2021). Today, the image of a *Jezebel* is one of a promiscuous and manipulative seductress (Ladson-Billings 2009: 88), while being an attractive Black, light-skinned or T woman. This might be illustrated, by way of example, by the actresses Dorothy Dandridge, Lena Horn and Halle Berry being cast in the role of a *Jezebel* (Ladson-Billings 2009: 89).

In *How To Get Away With Murder,* Professor Keating is portrayed as an excessively aggressive Black woman, furthering her career by seducing men and taking advantage of them. She is depicted as "unfit for motherhood", "uncontrollably sex-crazed" and "lacking all moral grounding" (Gammage 2015: 118). For instance, Professor Keating uses her affair with Nate to illegally obtain evidence to win cases for her clients and to protect her students.

Also, Professor Keating's image in the show might be characterized as the one of the *Angry Black Woman* in that she is depicted as aggressive, masculinized, lacking compassion and kindness. This image is introduced right at the beginning of the show, where Professor Keating is cold-calling her students and behaves rather aggressively towards two of them – Wes and Laurel – thus setting the tone for the rest of show by remaining cold and aggressive in the classroom, in personal life, and in her work as an attorney. The image of the *Angry Black Woman* is highlighted by the tone of Professor Keating's voice, which is stern and angry throughout the show.

4 Conclusions

The show obviously aims at eliminating racial stereotypes, taking the focus away from color by "normalizing non-racialized characters [which] exemplifies the liberal individualist discourse of a post-racial America" (Warner 2014: 637).

However, the portrayal of Professor Keating is still founded on Black stereotypes and on the negative image of criminal defense attorneys which might produce negative sentiments and form fallacious ideas of an Afro-American attorney in viewers' minds, thus leading to their reliance on that distorted information. Indeed, the negative image of Professor Keating as a Black female attorney could contribute to a negative bias and, therefore, to difficulties for Black women in achieving success in law firms.

Also, in a sense, the show provides a distorted image of university life and law students' routines. While it appears quite comprehensible that watching a show depicting the realities of studying at a law school would be dull and boring for viewers, popularization of legal education and criminal process in *How To Get Away With Murder* might contribute to the formation of unrealistic expectations in teenagers and aspiring lawyers, and shape a skewed and distorted perception of the American criminal system. Indeed, studies show that many viewers tend to think that TV shows and legal dramas are an accurate portrayal of the criminal justice system at work, and that those shows and legal dramas are lessons in law transmitted directly into their living rooms (Stockwell 2005).

Books, television, and other forms of mass media have always influenced people's perceptions, in particular those of the criminal justice system and legal profession. However, few have had such an effect and public interest as have recent legal dramas, one of the prominent examples of that being *How To Get Away With Murder.*

References

Brown Graves, Sheryl. 1999. Television and prejudice reduction: When does television as a vicarious experience make a difference, *Journal of Social Issues* 55 (4). 707-727.
Enlow, Courtney. 2015. Shonda Rhimes doesn't want to answer your questions about Diversity anymore. https://www.vh1.com/news/o2ch73/shonda-rhimes-diversity (accessed 13 May 2022).

Essed, Philomena. 1990. Everyday racism: Reports from women of two cultures. Alameda: Hunter House.

Gammage, Marquita Marie. 2015. Representations of Black Women in the Media: The Damnation of Black Womanhood. New York: Routledge.

Gerbner, George, Larry Gross, Nancy Signorelli, Michael Morgan. 1980. Aging with television: Images on television drama and conceptions of social reality. *Journal of Communication* 30. 37-48.

Heinrich Böll Stiftung. 2010. Rassismus & Diskriminierung in Deutschland. Dossier. https://heimatkunde.boell.de/sites/default/files/dossier_rassismus_und_diskriminierung.pdf

Howard, Annie. 2015. Viola Davis on 'How to Get Away with Murder' Role: "A Real Woman on TV in the Middle of This Pop Fiction". https://www.hollywoodreporter.com/news/general-news/viola-davis-how-get-away-801680/ (accessed 8 May 2022).

Hurwitz, Jon, Mark Peffley. 1997. Public perceptions of race and crime: the role of racial stereotypes. *American Journal of Political Science* 41(2). 375-401.

Jewell, Sue K. 1993. From mammy to miss America and beyond: Cultural images and the shaping of US policy. New York: Routledge.

Ladson-Billings, Gloria. 2009. "Who you callin' nappy-headed?" A critical race theory look at the construction of Black women. *Race Ethnicity and Education* 12:1. 87-99.

Martens, Emiel, Debora Povoa. 2017. How to get away with color: color-blindness and the post-racial illusion in popular American television series. *Alphaville Journal of Film and Screen Media* 13. 117-134.

Nilsen, Sarah, Sarah E. Turner. 2014. The colorblind screen: Television in post-racial America. London/ New York: New York University Press.

Pettigrew, Thomas Fraser, Roel W. Meertens. 1995. Subtle and blatant prejudice in Western Europe. *European Journal of Social Psychology* 25 (1). 55-75.

Prudom, Laura. 2014. "How to get away with murder" creator Peter Nowalk on working with Shonda Rhimes, *Diversity on TV*. https://variety.com/2014/tv/news/how-to-get-away-with-murder-creator-peter-nowalk-shonda-rhimes-viola-davis-diversity-1201313779/ (accessed 13 May 2022).

Rodriguez, Ana Eckhardt. 2019. The Black gangster and the Latino cleaning lady. Discrimination and Prejudices on Television. *Televizion* 32. 22-25.

Scott, C. Ellen. 2016. Cinema Civil Rights: Regulation, Repression, and Race in the Classical Hollywood Cinema. Rutgers University Press.

Sherwin, Richard Kenneth. 2006. Popular culture and law. *International Library of Essays in Law and Society, NYLS Legal Studies Research Paper* 07/08-4. https://papers.ssrn.com/sol3/papers.cfm?abstract_id=1004740# (accessed 10 May 2022).

Shohat, Ella, Robert Stam. 2014. Unthinking Eurocentrism: Multiculturalism and the Media. London: Routledge.

Stockwell, Jamie. 2005. More juries taking TV to heart. *Houston Chronicle* A2.

Supiano, Beckie. 2015. Racial disparities in higher education: An overview. *The Chronicle of Higher Education*. www.chronicle.com/article/Racial-Disparities-in-Higher/234129 (accessed 12 May 2022).

Tan, Alexis S., Gerdean Tan. 1979. Television use and self-esteem of Blacks. *Journal of Communication* 29. 129-135.

Torres-Spelliscy, Ciara, Monique Chase & Emma Greenman. 2010. Improving judicial diversity. *Brennan Center for Justice, New York University School of Law*. www.brennancenter.org/sites/default/files/legacy/Improving_Judicial_Diversity_2010.pdf (accessed 9 May 2022).

Turchiano, Danielle. 2020. 'How To Get Away With Murder' team reflects on road to series finale. https://variety.com/2020/tv/features/how-to-get-away-with-murder-series-finale-preview-viola-davis-interview-cliffhanger-1203538772/ (accessed 8 May 2022).

Van Dijk, Teun A. 2012. The role of the press in the reproduction of racism. In Michi Messer, Renee Schroeder & Ruth Wodak (Eds.), *Migrations: interdisciplinary perspectives*. Wien: Springer. 15-29.

Warner, J. Kristen. 2014. The racial logic of Grey's anatomy: Shonda Rhimes and her 'post-civil rights, post-feminist' series. *Television & New Media* vol. 16, no. 7.

Websites

Britannica. 2021. Jezebel. Queen of Israel. https://www.britannica.com/biography/Jezebel-queen-of-Israel (accessed 9 May 2022).

DataUSA. 2022. University of Southern California. https://datausa.io/profile/university/university-of-southern-california (accessed 12 May 2022).

Penn Law. 2014. Biddle Plus: How to Get Away with Murder. https://www.law.upenn.edu/live/news/5082-biddle-plus-how-to-get-away-with-murder (accessed 13 May 2022).

Philadelphia Inquirer. 2014. Penn law students say "Murder" gets away with murder. https://www.inquirer.com/philly/entertainment/television/20141007_Penn_law_students_say__Murder__gets_away_with_murder.html (accessed 13 May 2022).

Statista. 2018. Average daily on-demand TV and video viewing time in selected countries worldwide. https://www.statista.com/statistics/276748/average-daily-tv-viewing-time-per-person-in-selected-countries/ (accessed 10 May 2022)

Statista. 2022. Average daily time spent watching TV in the United States from 2019 to 2023. https://www.statista.com/statistics/186833/average-television-use-per-person-in-the-us-since-2002/ (accessed 10 May 2022)

Statista. 2022. How many hours of online video do you watch per week? https://www.statista.com/statistics/611707/online-video-time-spent/ (accessed 10 May 2022).

Statista. 2022. Share of people watching content through a streaming service in the United States in 2021, by hours per week. https://www.statista.com/statistics/819483/time-spent-streaming-video/ (accessed 10 May 2022).

Texas Penal Code. 2019. https://statutes.capitol.texas.gov/Docs/PE/htm/PE.38.htm#38.122 (accessed 4 September 2021).

About the authors

Patrizia Anesa, PhD, is Associate Professor in English Language and Translation at the University of Bergamo, Italy. She holds a Ph.D. in English Studies, with a specialization in professional communication. She is a member of the Research Centre on Specialized Languages (CERLIS), an Associate Editor of the IDEA project (International Dialects of English Archive), and currently the editor-in-chief of The International Journal of Law, Language, and Discourse. Her research interests lie mostly in the area of specialized discourse. In particular, she is currently interested in the applications of Conversation Analysis in LSP and the investigation of knowledge asymmetries in expert-lay communication.

Email: patrizia.anesa@unibg.it

Farida Buniatova holds a PhD in Law, with a specialisation in Civil and Private International Law. She is currently enrolled in the LLM program in Fashion Law at the Università Cattolica del Sacro Cuore of Milan. She is a member of the European Legal Interpreters and Translators Association (EULITA) and of the European Law Institute (ELI). Her research interests lie mostly in the areas of legal translation and cross-cultural legal communication, as well as in family and succession laws.

Email: faridabouniatova@hotmail.com

Giuliana Diani is Associate Professor of English Language and Translation at the University of Modena and Reggio Emilia, Italy. She holds an MA in Language Studies from the University of Lancaster (UK) and a Ph.D. in English Linguistics from the University of Pisa (Italy). She has worked and published on various aspects of discourse analysis and EAP, with special reference to language variation across academic genres, disciplines and cultures through the analysis of small, specialised corpora. Her recent work has centred on the analysis of the popularisation strategies adopted in print and in digital texts targeting young people and in law blogs.

Email: giuliana.diani@unimore.it

Jan Engberg is Professor of Knowledge Communication at the School of Communication and Culture, Section of German Business Communication, University of Aarhus, Denmark. His main research interests are the study of cognitive aspects of specialized discourse and the relation between specialized knowledge and text formulation. Much of his research is focused upon communication, translation and meaning in the field of law. However, especially with Carmen Daniela Maier he has also studied academic publishing as well as disseminating genres from a multimodal point of view. He has published widely in the field and co-edited a number of books and special issues of international journals. Finally, he is co-editor of the international journals *Fachsprache – Journal of Professional and Scientific Communication* and *Hermes – Journal of Language and Communication for Business.*

Email: je@cc.au.dk

Laura Clemenzi, PhD, is a Research Fellow at the Tuscia University (Viterbo, Italy), where she teaches Italian Linguistics. In 2015 she obtained a PhD in Italian Linguistics at Sapienza University of Rome with a dissertation on the language of the Italian industrial documentaries. During her studies she focused on terminology and corpus linguistics at the Pompeu Fabra University in Barcelona. Her main interests and published works concern the specialized lexicon, the legal and media discourse, the representation of immigration in cinema and the didactic of Italian as foreign and second language. Currently she is involved in the national Italian projects *Clarity in Court Proceedings (ClarAct): a new*

https://doi.org/10.1515/9783111048789-010

database for scholars and citizens and *Geography and History of Italian Grammars (GeoStoGrammIt)*, and she collaborates with the Accademia della Crusca for its Linguistic Advice service.
Email: laura.clemenzi@unitus.it

Francesca Fusco, PhD, is a Research Fellow at the University of Padua in the framework of the ERC-Consolidator Grant 2020 *MICOLL: Migrating Commercial Law and Language. Rethinking Lex Mercatoria (11th-17th Century)*. She holds a Master Degree in Law (Bocconi University, 2010), a Bachelor Degree in Foreign Languages and Literature, a Master Degree in Italian Language and Literature (University of Milan, 2014 and 2016) and a PhD in Linguistics (Sapienza University of Rome, 2020). She teaches *Writing Techniques for Journalism* at LUMSA University of Rome. She is a member of the editorial board of the journal *Testo e Senso* and of the scientific committee of *Commissione Lingua e Diritto* of the *Unione Nazionale Camere Civili*. From 2020 to 2022 she participated as a Research Fellow to the national Italian project *Clarity in Court Proceedings (ClarAct): a new database for scholars and citizens*. In 2013 she passed the bar exam. Her research interests mainly concern the language of law and administration (from a synchronic and diachronic point of view) and lexicography.
Email: francesca.fusco@unipd.it

Daniele Fusi is a Digital Humanities scholar mainly focused on digital text, and a professional full-stack software developer with decades of experience in specialized software solutions for both Academic institutions and business companies, especially in the publishing field. He combines these skills with his education in Classics, mainly focused in Greek and Latin language, philology and metrics, digital humanities, historical linguistics, lexicography, and epigraphy. After his degree in Classics at the Sapienza University of Rome, he got a PhD in Greek and Latin Philology at the University of Genoa. Since then, he has been working for many research projects both as an IT consultant and developer and as a scholar, while also teaching DH for many years at the Sapienza University of Rome and University of Macerata. In 2019 he entered Venice VeDPH as a Research Fellow in Digital Textual Scholarship, and in 2021 had a 1-year tenure at Bamberg as *Professur für Geschichte und Kultur der Spätantike* in the context of an ERC project. Many of his recent projects are published in the form of open source code at the VeDPH GitHub repository and at his own site (https://github.com/vedph, https://myrmex.github.io/overview).
Email: daniele.fusi@unive.it

Daniel Greineder is an international arbitration practitioner and was called to the Bar of England and Wales in 2005. He has practised in London, at a leading international law firm, as well as in Geneva, at a Swiss-Korean law firm, and most recently in Doha. He routinely works with lawyers from different common and civil law backgrounds, and many of his cases involve the application of foreign law. In this, he benefits from being a native speaker of both English and German. His work is recognized in leading directories, such as *Who's Who Legal*, and he is regular contributor to arbitral debate both as a conference speaker and in his publications.

Before qualifying as a barrister, he completed a doctorate on the role of mythology in late eighteenth-century German literary theory at the University of Oxford and maintains an interest in cultural, linguistic and literary matters.
Email: danielgreineder@aol.com

Mary C. Lavissière is an Associate Professor of Applied Languages at Nantes Université, France. Trained in English and Spanish linguistics, she focuses on language for specific purposes, linguistically describing and teaching about legal genres, morphosyntax, historical linguistics and

linguistic methods applied to discourse in management sciences. Her recent work includes projects on legal genres, modernisation of legal language, textometry applied to management sciences, and linguistic traits of women in male-dominated institutions, such as the maritime industry.
Email: marycatherine.lavissiere@univ-nantes.fr

Alessandra Lombardi, PhD, is Associate Professor in German Language and Translation at the Catholic University of Brescia, Italy. She holds a PhD in Translation Studies, with a specialization in legal translation and legal linguistics. She is a member of the Research Centre on Specialized Languages (CERLIS) and co-founder of the Italy-Germany Networking Initiatives for Art Communication and Promotion (IGENI). Main research interests: contrastive linguistics (Italian-German), LSP (text-)linguistics and LSP translation (focus on academic, legal and tourist texts), corpus-based terminology and terminography.
Email: alessandra.lombardi@unicatt.it

Giulia Lombardi is a Research Fellow in Italian Linguistics at the University of Genoa, where she is involved in the national Italian project *Clarity in Court Proceedings (ClarAct): a new database for scholars and citizens*. She holds a degree in Classics and a PhD in Digital Humanities. She teaches *Teaching Italian for foreigners* at the University of Genoa and she is a member of the university group for the evaluation of students' incoming competences. She is a member of the board of the AQuAA Association (Association for the Quality of Administrative Acts) founded by the ITTIG-CNR and the Accademia della Crusca. Her research interests focus mainly on legal linguistics, second language acquisition and pragmatics.
Email: giulia.lombardi@edu.unige.it

Karin Luttermann is Professor of German Linguistics at the Department of Linguistics and Literature at the University of Eichstaett-Ingolstadt, Germany. Topic of her habilitation thesis: Sprachgebrauch und Verständlichkeit. Member of the Graduate School of the Faculty of Linguistics and Literature at the University of Milan, Italy. Member of the Society for Applied Linguistics (GAL), member of the advisory board of the Austrian Association for Legal Linguistics (ÖGRL) and of the journal Comparative Law and Language (CLL). Co-editor of the book series Legal Linguistics, Münster (Lit). Main research interests: Expert-lay communication, multimodal communication, popularisation, comprehensibility in the field of law.
Email: karin.luttermann@ku.de

Jekaterina Nikitina is a Research Fellow in English Language and Translation at the University of Milan, where she lectures in linguistic mediation and discursive practices in legal and international settings. She holds a PhD in Linguistic, literary and intercultural studies in European and extra-European perspectives from the University of Milan. Her academic interests include specialised discourse and translation studies. She works on LSP theories and applications, knowledge dissemination dynamics, legal discourse, discourse of healthcare, medicine and bioethics. In her research, she applies qualitative and quantitative, specifically corpus linguistics, analytical approaches. Her published academic work includes publications on legal translation, discourse of human rights, dissemination of bioethical knowledge, popularisation of scientific topics through a range of channels, with a focus on medically assisted procreation and gene editing. She is a member of the Corpus and Language Variation in English Research Group (CLAVIER) and of the Italian Association of Translators and Interpreters (AITI).
Email: Jekaterina.nikitina@unimi.it

Katia Peruzzo is Research Fellow in English Language and Translation at the Department of Legal, Language, Interpreting and Translation Studies of the University of Trieste, Italy. She holds a PhD in Translation and Interpreting Studies and is a member of the Corpus and LAnguage Variation In English Research group (CLAVIER). Her longer-standing research interest is in legal discourse, with emphasis on legal terminology in the criminal field and legal and judicial translation for international audiences. More recently, her studies have geared towards knowledge dissemination and popularisation, with her latest papers applying mainly qualitative corpus linguistics analytical approaches to focus on the mediation of legal knowledge for children and young audiences.
Email: kperuzzo@units.it

Federica Scarpa is Full Professor of English Language and Translation at the Department of Legal, Language, Interpreting and Translation Studies of the University of Trieste, Italy. She teaches Translation and Specialized Translation from English into Italian. She has been the Director of the Master in Legal Translation and Coordinator of the PhD Programme in Interpreting and Translation Studies. She has published three volumes and numerous articles and contributions on technical and scientific translation as well as legal translation.
Email: fscarpa@units.it

Dieter Stein is emeritus professor of English linguistics at Heinrich-Heine-University Düsseldorf. He continues to teach at this and other universities in Europe, China and the Americas. His main research interests and publications are in the field of language change, pragmatics, communication in the digital media, as well as language in the law and forensic linguistic. He founded several Open Access e-journals ("Language in the internet" and "The International Journal of language and law") and served as editor-in-chief of LSA's eLanguage digital publication portal. He was president of the International Society for Historical Linguistics and of the International Law and Language Association (ILLA).
Email: stein@hhu.de

Index

https://doi.org/10.1515/9783111048789-011

www.ingramcontent.com/pod-product-compliance
Lightning Source LLC
Chambersburg PA
CBHW051426090426
42737CB00014B/2842